The Practical Approach Series

SERIES EDITORS

D. RICKWOOD
*Department of Biology, University of Essex
Wivenhoe Park, Colchester, Essex CO4 3SQ, UK*

B. D. HAMES
*Department of Biochemistry and Molecular Biology
University of Leeds, Leeds LS2 9JT, UK*

Affinity Chromatography
Anaerobic Microbiology
Animal Cell Culture
 (2nd Edition)
Animal Virus Pathogenesis
Antibodies I and II
Basic Cell Culture
Behavioural Neuroscience
Biochemical Toxicology
Biological Data Analysis
Biological Membranes
Biomechanics—Materials
Biomechanics—Structures and
 Systems
Biosensors
Carbohydrate Analysis
 (2nd Edition)
Cell–Cell Interactions
The Cell Cycle
Cell Growth and Division
Cellular Calcium
Cellular Interactions in
 Development
Cellular Neurobiology

Centrifugation (2nd Edition)
Clinical Immunology
Computers in Microbiology
Crystallization of Nucleic Acids
 and Proteins
Cytokines
The Cytoskeleton
Diagnostic Molecular Pathology
 I and II
Directed Mutagenesis
DNA Cloning I, II, and III
Drosophila
Electron Microscopy in Biology
Electron Microscopy in
 Molecular Biology
Electrophysiology
Enzyme Assays
Essential Developmental
 Biology
Essential Molecular Biology I
 and II
Experimental Neuroanatomy
Extracellular Matrix
Fermentation

Plant Cell Culture

A Practical Approach
Second Edition

Edited by

RICHARD A. DIXON

and

ROBERT A. GONZALES

Plant Biology Division,
The Samuel Roberts Noble Foundation,
P.O. Box 2180,
Ardmore, Oklahoma 73402, USA

IRL PRESS
at
OXFORD UNIVERSITY PRESS
Oxford New York Tokyo

Oxford University Press, Great Clarendon Street, Oxford OX2 6DP

Oxford   New York
Athens   Auckland   Bangkok   Bombay
Calcutta   Cape Town   Dar es Salaam   Delhi
Florence   Hong Kong   Istanbul   Karachi
Kuala Lumpur   Madras   Madrid   Melbourne
Mexico City   Nairobi   Paris   Singapore
Taipei   Tokyo   Toronto

and associated companies in
Berlin   Ibadan

Oxford is a trade mark of Oxford University Press

Published in the United States by
Oxford University Press Inc., New York

© Oxford University Press, 1994

First published 1994
Reprinted 1996

A catalogue record for this book is available from the British Library

Library of Congress Cataloging-in-Publication Data
Data available

ISBN 0 19 963403 3 (Hbk)
ISBN 0 19 963402 5 (Pbk)

Printed in Great Britain by
Information Press Ltd, Eynsham, Oxford

Preface

The second edition of this book incorporates a number of significant changes to the content and scope of the first edition (published in 1985). These primarily reflect the impact of molecular biology on the plant sciences. Regeneration of plants from cell culture is a key step in many genetic transformation strategies, and protoplasts have increasingly been used as recipient cells for DNA delivery for both transient gene expression and stable transformation studies. The last few years have seen rapid progress in understanding plant stress physiology at the molecular level, hence the inclusion of new sections on cell selection for cold, salinity, and toxin resistance. Whereas in 1985 the manipulation of secondary metabolite levels in cell cultures could only be performed by varying the chemical and physical parameters of the culture environment, today both levels and composition of secondary metabolites can be manipulated by gene transfer. We hope the new sections will be of value to plant molecular biologists, particularly by providing specific examples of the different tissue culture systems and approaches that can be used. We have, however, omitted details of recombinant DNA techniques from this volume, as these are readily available from other volumes in this series.

As in the first edition, this volume describes methods for the establishment and manipulation of well-characterized cell culture systems, as well as more general advice that should enable researchers to develop their own protocols with their specific species of interest. It is still true to say that development of tissue culture protocols is in part an empirical science (or perhaps even an 'art'), but the increasing body of information available makes it ever easier to make generalizations that will have some utility. We hope that the scope of this second edition will adequately encompass the broad range of areas, both fundamental and applied, in which plant cell culture technology is a central component.

Ardmore   R. A. D.
November 1993   R. A. G.

Contents

Contents

3. Applications of protoplast technology   41

Contributors

ERICA E. BENSON
Department of Molecular and Life Sciences, The University of Abertay Dundee, Bell Street, Dundee DD1 1HG, Scotland, UK.

N. W. BLACKHALL
Department of Life Science, University of Nottingham, University Park, Nottingham NG7 2RD, UK.

JAN BRAZOLOT
Department of Crop Science, Ontario Agricultural College, University of Guelph, Guelph, Ontario N1G 2W1, Canada.

D. C. W. BROWN
Plant Research Center, Central Experimental Farm, Agriculture Canada, Ottawa, Ontario K1A 0C6, Canada.

M. R. DAVEY
Department of Life Science, University of Nottingham, University Park, Nottingham NG7 2RD, UK.

PIERRE C. DEBERGH
Laboratory of Horticulture, State University Gent, Faculty of Agricultural Sciences, Coupure Links 653, B-9000 Gent, Belgium.

R. A. DIXON
Plant Biology Division, Samuel Roberts Noble Foundation, Inc., P.O. Box 2180, Ardmore, Oklahoma 73402, USA.

JOHN J. FINER
Department of Agronomy, Ohio State University, 1680 Madison Avenue, Wooster, Ohio 44691, USA.

C. I. FRANKLIN
Plant Biology Division, Samuel Roberts Noble Foundation, Inc., PO Box 2180, Ardmore, Oklahoma 73402, USA.

ROBERT A. GONZALES
Plant Biology Division, Samuel Roberts Noble Foundation, Inc., PO Box 2180, Ardmore, Oklahoma 73402, USA.

K. PETER PAULS
Department of Crop Science, Ontario Agricultural College, University of Guelph, Guelph, Ontario N1G 2W1, Canada.

J. B. POWER
Department of Life Science, University of Nottingham, University Park, Nottingham NG7 2RD, UK.

JAMES C. REGISTER III
Pioneer Hi-Bred International, Inc., Department of Analytical Biochemistry, 7300 NW 62nd Ave., PO Box 1004, Johnston, Iowa 50131, USA.

RICHARD J. ROBINS
Genetics and Microbiology Department, Institute of Food Research (Norwich Laboratory), Norwich Research Park, Colney, Norwich NR4 7UA, UK.

A. H. SCRAGG
Department of Molecular Biology, University of the West of England, Coldharbour Lane, Frenchay, Bristol BS16 1QY, UK.

JOHN L. SHERWOOD
Department of Plant Pathology, Oklahoma State University, Stillwater, Oklahoma 74078, USA.

J. SINGH
Plant Research Center, Central Experimental Farm, Agriculture Canada, Ottawa, Ontario K1A OC6, Canada.

STEFAAN P. O. WERBROUCK
Laboratory of Horticulture, State University Gent, Faculty of Agricultural Sciences, Coupure Links 653, B-9000 Gent, Belgium.

I. WINICOV
Departments of Microbiology and Biochemistry, University of Nevada, Reno, Nevada 89557, USA.

KANG FU YU
Department of Crop Science, Ontario Agricultural College, University of Guelph, Guelph, Ontario N1G 2W1, Canada.

Abbreviations

ABA	abscisic acid
ADC	arginine decarboxylase
B5	Gamborg's B5 medium
BAP	6-benzylaminopurine
BSA	bovine serum albumin
CaMV	cauliflower mosaic virus
CAT	chloramphenicol acetyltransferase
CP	coat protein
CPMP	coat protein-mediated protection
2,4-D	2,4-dichlorophenoxyacetic acid
DC	direct current
DFMA	α-difluoromethyl-DL-arginine
DFMO	α-difluoromethyl-DL-ornithine
DM	dry mass
DMSO	dimethylsulfoxide
EDTA	ethylenediamine tetraacetic acid
FDA	fluorescein diacetate
FID	flame ionization detector
FM	fresh mass
GA	gibberellic acid
GC	gas chromatography
GUS	β-glucuronidase
Hepes	N-2-hydroxyethylpiperazine-N'-2-ethanesulfonic acid
HPLC	high performance liquid chromatography
I_{50}	concentration resulting in 50% inhibition
IAA	indole acetic acid
IBA	indole-3-butyric acid
2iP	isopentenyladenine
K	kinetin (6-furfurylaminopurine)
LDC	lysine decarboxylase
MeOH	methanol
MES	2-(N-morpholino)-ethanesulfonic acid
MIC	minimum concentration resulting in 100% inhibition
MS medium	Murashige and Skoog medium
NAA	1-naphthaleneacetic acid
NMR	nuclear magnetic resonance
NOA	naphthoxyacetic acid
ODC	ornithine decarboxylase
PBS	phosphate-buffered saline

pCPA	*p*-chlorophenoxyacetic acid
PCR	polymerase chain reaction
PCV	packed cell volume
PDA	potato dextrose agar
PEG	polyethylene glycol
PGR	plant growth regulator
PMT	putresceine *N*-methyltransferase
PND	phosphorus/nitrogen detector
PVDF	polyvinylidene difluoride
RH	relative humidity
SH medium	Schenk and Hildebrandt medium
STR	stirred-tank bioreactor
TDC	tryptophan decarboxylase
TDZ	thidiazuron
TMV	tobacco mosaic virus
TTC	triphenyl tetrazolium chloride
Zea	zeatin
ZR	zeatin riboside

1

Initiation and maintenance of callus and cell suspension cultures

C. I. FRANKLIN and R. A. DIXON

1. Introduction

Establishing dedifferentiated cultures from organized plant tissues was a major goal of early plant cell culture studies. Today, this is a relatively routine procedure which is an essential prerequisite for a range of subsequent approaches including regeneration, embryogenesis, growth of large scale cultures, and selection strategies. The key to establishing callus and cell suspension cultures is the choice of the optimum culture medium components, proper explant source, and plant growth regulator concentrations. This is still largely empirical. However, a large number of plant species from a wide-range of families have now been successfully grown in culture, and access to this information allows the researcher to make a good guess at conditions which may be suitable for the growth of species whose culture requirements have not been determined.

In the first edition of this book (1), we discussed in some detail the basic equipment and facilities necessary for media preparation and the initiation and growth of plant cell cultures. These have also been described in other more comprehensive reference books (2, 3). The following list outlines the major items you will require:

- chemical balance
- pH meter
- volumetric flasks and beakers for media preparation
- culture vessels (Petri dishes and conical flasks)
- autoclave
- laminar flow hood for aseptic transfers
- open platform orbital shakers (such as those sold by L. H. Engineering, UK, or New Brunswick Scientific, USA)
- constant temperature room or incubator(s) which can be set at 25 ± 2 °C
- fluorescent lighting controlled by a timer

It is possible to dispense with the sensitive chemical balance if pre-made culture media are purchased, and to work without a laminar flow hood if a clean area of laboratory can be set aside, and stringent aseptic technique applied. However, microbial contamination of cultures is difficult to avoid completely if a transfer hood is not used.

2. Culture media

The first decision to be made when initiating a plant callus culture is the composition of the culture medium. A large number of different culture media are described in the literature, but of these, only a few have found widespread usage for a range of plant species (e.g. MS, SH, and B5, see *Table 1*). In a culture medium containing 20 or more components, there is an infinite possibility for variation of composition. However, there is sufficient informa-tion in the literature to indicate which features of the culture medium are important, requiring optimization, and which, within certain prescribed limits, are of less importance. *Table 2* presents a very selective overview of plant species which have been successfully grown in callus or suspension culture, and provides details of the basic medium and growth regulator concentrations. For an extensive list of media formulations for various plant species refer to George *et al.* (12). If you wish to work with a species listed in *Table 2*, it would be wise to start with the culture conditions referred to therein. However, remember that there may be significant differences be-tween the behaviour of individual cultivars of a species with respect to growth in culture, and that some optimization of culture conditions may be necessary (see Section 2.2). If your species is not listed in *Table 2*, or George *et al.* (12), decide on your initial test medium by comparing media used for closely related species.

2.1 Composition of commonly used culture media

This and the following section outline the components found in plant cell culture media. Methods for preparing culture media are discussed in Section 2.3. Probably the most commonly used plant culture medium is that of Murashige and Skoog (4) (*Table 1*). As with nearly all plant culture media, its components can be divided into six groups. These are:

- major inorganic nutrients
- trace elements
- iron source
- vitamins
- carbon source
- plant growth regulators

These divisions are both functional and practical (i.e. they reflect the way in which the various stock solutions are made up). Note that plant growth regulator concentrations are omitted from the basic medium formulation. This is because, by alteration of growth regulator concentrations, it is possible to greatly affect growth and differentiation in culture. Growth regulator concentrations required for maintenance of a dedifferentiated callus line of one species may induce organogenesis in another species. The variations in growth regulator levels required for callus growth of different species are clear from a brief inspection of *Table 2*. It is important to remember that reference to a culture medium by its abbreviation (e.g. MS, SH) generally implies the culture medium minus growth regulators. These are listed in full after the medium name, e.g. MS medium supplemented with 10^{-6} M 2,4-D and 5×10^{-7} M kinetin. More detailed advice on choice of plant growth regulators is given in Section 2.2.

Components of a few selected tissue culture media are compared in *Table 1*. One of the important components of the basal medium is nitrogen. MS medium has high levels of inorganic nitrogen while other formulations, such as NN, have very low levels. Some species may require or tolerate higher levels of nitrogen in the medium than others. Certain explants of some species do not require or can not tolerate higher levels of nitrogen in the medium, e.g. immature ovules of *Impatiens platypetala* (81). With some species, e.g. rice, the ratio between NH_4^+ and NO_3^- is very critical for cell culture and plant regeneration (42). In general, there is a tendency to use lower levels of NH_4^+ than NO_3^- in plant tissue culture media. The amount of NH_4^+ in MS medium is almost half that of NO_3^-, while other formulations, such as SH, B5, and GD, contain even lower levels of NH_4^+.

There are several reports in the literature describing the influence of organic nitrogen in the tissue culture medium. Organic nitrogen has been shown to enhance somatic embryogenesis in some species, e.g. orchard grass (56), *Agrostis* (57), and organogenesis in others, e.g. green bean (11). Most commonly, inorganic nitrogen in the medium is supplemented with organic nitrogen in the form of proline or glutamine (57–59). In a few tissue culture systems the organic nitrogen (e.g. glutamine) serves as the sole nitrogen source in the medium (11). The components of the SI medium used for regenerating shoots from bean cotyledons (11) are identical to those of MS medium except for the nitrogen source and sucrose concentration (*Table 1*). The inorganic nitrogen is completely replaced by 10 mM glutamine (1462 mg/litre) which is filter sterilized and added to the autoclaved and cooled medium. Deletion of KNO_3 lowers the K^+ concentration in the medium by 18.8 mM, but this can be compensated for by adding 18.8 mM (1400 mg/litre) KCl to the medium. We have not observed any harmful effects of elevated Cl^- levels in our bean tissue culture system.

The LM medium (9) listed in *Table 1* is ideal for initiation and maintenance of rapidly growing loblolly pine cell cultures (C. I. Franklin, unpublished

3

Table 1. A comparison of components in commonly used plant tissue culture media or media modified for a particular species or for a specific purpose (see text)

Constituent	Concentration in culture medium [b] (mg/litre)							
	MS	SH	B5	GD	KM [a]	LM	NN	SI
Ca(NO₃)₂				241				
KNO₃	1900	2500	2500	1000	1900	1900	950	
NH₄NO₃	1650			1000	600	1650	720	
NH₄H₂PO₄		300						
(NH₄)₂SO₄			134					
MgSO₄·7H₂O	370	400	250	35	300	1850	185	370
CaCl₂·2H₂O	440	200	150		600	22	220	440
KCl				65	300			1400
KH₂PO₄	170			300	170	340	68	170
NaH₂PO₄·H₂O			150					
MnSO₄·H₂O		10.0	10.0	1.12	10.0	21		
MnSO₄·4H₂O	22.3						25	22.3
KI	0.83	1.0	0.75	0.8	0.75	4.15		0.83
H₃BO₃	6.2	5.0	3.0	0.3	3.00	31	10	6.2
ZnSO₄·7H₂O	8.6	1.0	2.0	0.3	2.00	43	10	8.6

	MS	SH	B5	GD	KM	LM	NN	SI
$CuSO_4 \cdot 5H_2O$	0.025	0.2	0.025	0.025	0.025	0.50	0.025	0.025
$Na_2MoO_4 \cdot 2H_2O$	0.25	0.1	0.25	0.025	0.25	1.25	0.25	0.25
$CoCl_2 \cdot 6H_2O$	0.025	0.1	0.025	0.025	0.025	0.125	0.025	0.025
$FeSO_4 \cdot 7H_2O$	27.8	15.0	27.8	27.8		27.8	27.8	27.8
Sequestrene 330 Fe					28			
Na_2EDTA	37.3	20.0	37.3	37.3		37.3	37.3	37.3
Nicotinic acid	0.5	5.0	1.0	0.1		0.5	5	0.5
Pyridoxine-HCl	0.5	0.5	1.0	0.1	1	0.1	0.5	0.5
Thiamine-HCl	0.1	5.0	10.0	1.0	1	0.1	0.5	0.1
D-Biotin				0.2	0.01		0.05	
Folic acid					0.4		0.5	
myo-Inositol	100	1000	100	10	100	100	100	100
Glycine	2.0			4	0.1		2	2.0
Glutamine						1462		
Sucrose	30 000	30 000	20 000	20 000	20 000	30 000	20 000	10 000

a Additional components of KM medium (mg/litre): nicotinamide 1; calcium D-pantothenate 1; folic acid 0.4; *p*-aminobenzoic acid 0.02; choline chloride 1; riboflavin 0.2; ascorbic acid 2; vitamin A 0.1; vitamin D_3 0.01; vitamin B_{12} 0.02; sodium pyruvate 20; citric acid 40; malic acid 40; fumaric acid 40; fructose, ribose, xylose, mannose, rhamnose, cellobiose, sorbitol, and mannitol, 250 each; adenine 10; guanine, thymine, uracil, hypoxanthine, and xanthine, 0.03 each; all 21 L-amino acids 0.1 each except glutamine 5.6, alanine 0.6, glutamic acid 0.6 and cysteine 0.2.
b MS = Murashige and Skoog (4), SH = Schenk and Hildebrandt (5), B5 = Gamborg *et al.* (6), GD = Gresshoff and Doy (7), KM = Kao and Michayluk (8), LM = Litvay *et al.* (9), NN = Nitsch and Nitsch (10), and SI = Franklin *et al.* (11).

Table 2. Media and growth regulator levels used for callus and cell cultures of selected species (for an extensive list of other plant species see ref. 12)

Species	Explant	Medium [a]	Growth regulators (μM)	Reference
Allium cepa	Stem	S	2,4-D 1.3	15
Arabidopsis thaliana	Seeds, stem, or leaf	S	2,4-D 4.5–9 + K 2.3	16
Arachis hypogea	Cotyledon	MS	Pic 2.1	17
Asparagus officinalis	Hypocotyl	LS	2,4-D 4.5 + K 1.5	18
Avena sativa	Leaf or tiller buds	MS	2,4-D 9 + BAP 4.4 + CM 10%	19
Aylostera pseudodeminuata	Vascular tissue	S	2,4-D 5 + K 5.1	20
Beta vulgaris	Hypocotyl	S	NAA 53.7 + K 4.7	21
Catharanthus roseus	Leaf	B5	2,4-D 9	22
Cattleya sp.	Meristem	S	NAA 0.5 + K 0.9 + CM 15%	23
Cinchona ledgeriana	Hypocotyl	B5	2,4-D 4.5 + K 0.9	24
Coffea arabica	Leaf	S	2,4-D 5 + K20	25
Crassula argentea	Leaf	MS	Zea 108.2 (or) 2iP 98.4	26
Cucumis sativus	Internode	MS	NAA 1 + BAP 0.1	27
Cymbidium sp.	Shoot apex	Wh	NAA 26.9 + CM 10%	28
Ephedra foliata	Female gametophyte	MS	2,4-D 4.5 + CM 10%	29
Euphorbia tirucalli	Stem	MS	2,4-D 4.5 + NAA 10.7	30
Ginko biloba	Pollen	S	2,4-D 2.71–27.12 + CM 10–40%	31
Glycine max	Cotyledon	MS	2,4-D 22.6–181	32
Gossypium hirsutum	Hypocotyl	S	NAA 10.7 + BAP 0.4 + AdS 217.2	33
Haworthia planifolia	Leaf	LS	2,4-D 0.9 + K 4.7	34
Hordeum vulgare	Root	B5	2,4-D 9.4	35

Species	Explant	Medium	Hormones	No.
Lemna gibba	Frond	S	2,4-D 45.3 + 2iP 4.9	36
Lithops lesliei	Leaf	MS	2,4-D 4.5 + K 46.5	37
Mammillaria elongata	Tubercle	S	2,4-D 27.1 + K 4.7–9.3 (or) 2iP 4.9–9.8	38
Mangifera indica	Nucellus	S	2,4-D 9	39
Medicago sativa	Leaf, hypocotyl, or cotyledon	S	2,4-D 4.5 + K 0.93 + Ad 7.4	40
Nicotiana tabacum	Leaf	MS (or) B5	2,4-D 4.5	41
Oryza sativa	Immature embryo	S	2,4-D 9	42
Pennisetum americanum	Inflorescence	S	2,4-D 11.3	43
Petunia hybrida	Leaf	S	2,4-D 9 + K 1.2	44
Phoenix dactylifera	Shoot-tips or embryo	S	2,4-D 440 + 2iP 14.5	45
Pinus taeda	Stem	LM	2,4-D 22.6	9
Pisum sativum	Root	S	2,4D 6.8 + NAA 0.5 + IAA 5.7 + K 1.2	46
Pteris vittata	Rhizome	S	2,4-D 9 + CM 10%	47
Saccharum sp.	Leaf	MS	2,4-D 2.3–13.6	48
Sorghum bicolor	Leaf	MS	2,4-D 9 + K 0.5	49
Theobroma cacao	Cotyledon	B5	2,4-D 4.5 + K 0.9	50
Triticum aestivum	Immature embryo	S	2,4-D 9	51
Vitis vinifera	Stem, leaf, or fruit	S	NAA 0.5 + K 0.9 + CM 15%	52
Yucca schidigera	Hypocotyl	MS	2,4-D 4.5	53
Zamia pumila	Embryo	B5	2,4-D 2.3 + K 4.7–18.6	54
Zea mays	Immature embryo	S	2,4-D 4.5–6.8	55

[a] MS = Murashige and Skoog medium (4); LS = Linsmaier and Skoog medium (13); B5 = Gamborg's B5 medium (6); Wh = White's medium (14); S = specific medium formulation or modified for the study reported; 2,4-D = 2,4-dichlorophenoxyacetic acid; NAA = 1-naphthaleneacetic acid; IAA = indole-3-acetic acid; K = kinetin; Ad = adenine; AdS = adenine sulfate; 2iP = N-isopentenylaminopurine; CM = coconut milk; Pic = picloram; BAP = 6-benzylaminopurine.

results) and may be suitable for other conifers as well. The GD and NN media are relatively 'low-salt' media. The GD medium (7) has been very useful for recovering loblolly pine plantlets from regenerated shoots (C. I. Franklin, unpublished results). The NN medium (10) is useful for regenerating haploid plants from pollen grains. The KM medium listed in *Table 1* (medium No. 6, in ref. 8) is very complex with several additives such as organic acids, amino acids, nucleic acid bases, etc., yet it is a defined medium, and it has been reported to support the growth of *Vicia hajastana* cells at low initial population density (25–50 cells/ml) (8). However, addition of undefined media components such as casein hydrolysate and coconut water (see below) was necessary for culturing *V. hajastana* cells at a much lower density of 1–2 cells/ml (medium No. 8, in ref. 8).

A comparison between the media in *Table 1* highlights several important features:

- all the media are fully defined (i.e. no added biological extracts)
- a chelated iron source (Fe-EDTA) is preferred (less easily depleted)
- MS and SH are 'high salt' media
- MS and SH use both ammonium and nitrate ions as source of nitrogen
- SH medium contains a very high level of *myo*-inositol
- sucrose is the preferred carbon source

Generally, use of fully-defined media is recommended, as the addition of biological supplements (casein hydrolysate, yeast extract, or coconut milk) introduces possible variation due to differences in the composition of different batches of additive. However, these additives may be beneficial and may supplement or replace the vitamins in certain cases, e.g. White's medium (14). Yeast extract and casein hydrolysate are readily available from most microbiological suppliers (e.g. Oxoid Ltd. in the UK, Sigma Chemical Company, or Difco Laboratories in the USA). You can purchase coconut milk from the suppliers listed above or prepare it according to *Protocol 1*.

Protocol 1. Preparation of coconut milk

1. Drain the milk from a large number of coconuts.[a]
2. Deproteinize by boiling for 10 min.
3. Filter through Whatman No. 1 filter paper.
4. Autoclave in small batches ($\leq$ 100 ml).
5. Store at $-70\,°C$ until required.
6. Thaw and add to culture medium at a final level of $\sim$10% (v/v).

[a] Use of a large number of coconuts minimizes variation. Check the milk from each coconut individually before pooling, and discard any which appears 'bad'.

2.2 Plant growth regulators

Most plant culture media contain an auxin and a cytokinin. These two classes of growth regulators, at concentrations generally around 10 μM, help maintain dedifferentiated cell growth and promote cell division respectively. The most commonly used plant growth regulators are listed in *Table 3*. 2,4-D is the most widely used synthetic auxin, especially for gramineous species. 2,4-D and NAA have largely replaced the naturally occurring auxin, IAA, in cell culture media, as the latter is readily oxidized by plant cells. Often, callus growth may require lower levels of auxin than needed for callus induction. Very low levels of auxin, or complete omission, may often induce organogenesis.

BAP and kinetin are the most commonly used cytokinins. Very few culture media employ GA_3 for initiation or growth of callus cultures. However, this growth regulator has been shown to be beneficial for the growth of potato cells in callus and suspension culture (61). New or less commonly used plant growth regulators and their properties are listed in *Table 4*. These compounds are mainly either synthetic analogues of native growth regulators such as IAA or zeatin, or were originally developed as herbicides. Some of these compounds are more potent than the native growth regulators.

Usually, manipulation of auxin and cytokinin levels will be successful in defining a growth regulator balance necessary for the required behaviour in culture. An experimental approach for such a manipulation is outlined in *Protocol 2*. If desired results are not obtained using commonly used growth regulators, it is worth trying some of the novel growth regulators listed in *Table 4*.

Most of the auxins and cytokinins commonly used in plant cell culture work are synthetic. IAA, GA_3, zeatin, and ABA are naturally occurring plant growth regulators. In the last several years, a number of new naturally occurring plant growth regulating compounds have been discovered. These have not yet found use in cell culture formulations, but the most interesting are listed below since they may prove useful in the future. It is worth noting that a number of them are phenolic compounds.

- dehydrodiconiferyl alcohol glycosides (cytokinin-like) (70, 71)
- chlorogenic acid (protects IAA from oxidation) (72)
- jasmonates (promote leaf senescence) (73)
- flavonoids (inhibit auxin transport) (74)
- various ethylene inhibitors
- various GA_3 biosynthesis inhibitors (62)

A wide-range of other plant secondary products have been listed in the earlier literature as potential naturally occurring growth regulators (75).

Much of the older plant cell culture literature listed concentrations of

Table 3. Commonly used plant growth regulators

Class	Name	Abbreviation	Comments [a]
Auxin	Indole-3-acetic acid	IAA	Use for callus induction at 10–30 µM. Lowering to 1–10 µM can stimulate organogenesis. Is inactivated by light and readily oxidized by plant cells. The synthetic auxins below have largely superceded IAA for tissue culture studies.
	Indole-3-butyric acid	IBA	Use for rooting shoots regenerated via organogensis. Either maintain at low concentration (1–50 µM) throughout rooting process, or expose to high concentration (100–250 µM) for 2–10 days and then transfer to hormone-free medium. Can also use as a dip for *in vitro* or *ex vitro* rooting of shoots.
	2,4-Dichlorophenoxyacetic acid	2,4-D	Most commonly used synthetic auxin for inducing callus and maintaining callus and suspension cells in dedifferentiated state. Usually used as sole auxin source (1–50 µM), or in combination with NAA.
	p-Chlorophenoxyacetic acid	pCPA	Similar to 2,4-D, but less commonly used.
	1-Naphthaleneacetic acid	NAA	Synthetic analogue of IAA. Commonly used, either as sole auxin source (2–20 µM for callus induction and growth of callus and suspension cultures; 0.2–2 µM for root induction), or, in combination with 2,4-D.

Cytokinin	6-Furfurylaminopurine (kinetin)	K	Often included in culture media for callus induction, growth of callus and cell suspensions, and induction of morphogenesis (1–20 μM). Higher concentrations (20–50 μM) can be used to induce the rapid multiplication of shoots, axillary/adventitious buds, or meristems.
	6-Benzylaminopurine	BAP	Included in culture media for callus induction, and growth of callus and cell suspensions (0.5–5.0 μM), and for induction of morphogenesis (1–10 μM). More commonly used than kinetin for inducing rapid multiplication of shoots, buds, or meristems (5–50 μM).
	N-Isopentenylaminopurine	2iP	Less commonly used than K or BAP for callus induction and growth (2–10 μM), induction of morphogenesis (10–25 μM), or multiplication of shoots, buds, or meristems (30–50 μM).
	Zeatin	Zea	Seldom used in callus or suspension media. Can be used for induction of morphogenesis (0.05–10 μM). Zea is thermolabile and must not be autoclaved.
Gibberellin	Gibberellin A$_3$	GA$_3$	Seldom used in callus or suspension medium (one exception being potato (59)). Can promote shoot growth when added to shoot induction medium at 0.03–14 μM. Also used to enhance development in embryo/ovule cultures (0.3–48 μM). GA$_3$ is thermolabile and must not be autoclaved.
Abscisic acid	Abscisic acid	ABA	Used at concentrations of 0.04–10 μM to prevent precocious germination, and promote normal development of somatic embryos (60).

[a] The concentrations given represent values taken from the literature for a range of plant species. Specific applications for species not described in the literature (first check *Table 2*) should be determined experimentally.

Table 4. Novel and/or less commonly used plant growth regulators for plant tissue culture

Compound	Abbreviation/ common name	M_r	Function/activity	Reference
Chlorocholine chloride [a]	CCC	158.1	Inhibitor of GA_3 biosynthesis	62
N-(2-Chloro-4-pyridyl)-N'-phenylurea [a]	4-CPPU	247.7	Cytokinin-like	63
3,6-Dichloroanisic acid	Dicamba	221.0	Auxin	56
2-Naphthoxyacetic acid	NOA	202.2	Auxin	64
Phenylacetic acid [a]	PAA	136.2	Auxin	65
4-Amino-3,5,6-trichloropicolinic acid	Picloram	241.5	Auxin-like	66
Thidiazuron [a]	TDZ	220.2	Cytokinin-like	67
2,4,5-Trichlorophenoxyacetic acid	2,4,5-T	255.2	Auxin	68
Zeatin riboside [a]	ZR	363.4	Cytokinin	69

[a] Filter sterilization recommended.

Protocol 2. Experimental approach to optimize growth regulator levels

1. Select one or more commonly used media formulations (e.g. MS, SH, or B5).

2. Optimize the growth regulator(s) requirements by setting up experiments using concentrations and ratios of auxin and cytokinin listed below.

3. Select the combination of explant source, medium, and growth regulator levels giving best results.

4. Further optimization (if necessary) can be done by investigating the effects of growth regulator levels in between the concentrations listed here or by using other auxins or cytokinins.

2,4-D (μM)	Kinetin (μM)				
	0	0.5	1	2.5	5
1	1/0	1/0.5	1/1	1/2.5	1/5
2.5	2.5/0	2.5/0.5	2.5/1	2.5/2.5	2.5/5
5	5/0	5/0.5	5/1	5/2.5	5/5
10	10/0	10/0.5	10/1	10/2.5	10/5
20	20/0	20/0.5	20/1	20/2.5	20/5

growth regulators in mg/litre (= parts per million). The modern convention is to express all concentrations as molarity. *Table 5* shows the conversion from mg/litre to μM for the commonly used plant growth regulators.

Table 5. Conversion of mg/litre to μM concentrations for commonly used plant growth regulators

Growth regulators	μM NAA	2,4-D	IAA	IBA	BAP	K	2iP	Zea
M_r	186.2	221.0	175.2	203.2	225.2	215.2	203.2	219.2
mg/litre								
0.0001	0.0005	0.0004	0.0005	0.0005	0.0004	0.0005	0.0005	0.0005
0.001	0.005	0.0045	0.006	0.005	0.004	0.005	0.005	0.005
0.005	0.027	0.023	0.028	0.025	0.022	0.023	0.025	0.023
0.01	0.054	0.045	0.057	0.049	0.044	0.046	0.049	0.046
0.05	0.27	0.226	0.285	0.246	0.222	0.232	0.246	0.228
0.10	0.54	0.452	0.570	0.492	0.444	0.465	0.492	0.456
0.25	1.34	1.13	1.43	1.23	1.11	1.16	1.23	1.14
0.50	2.69	2.26	2.85	2.46	2.22	2.32	2.46	2.28
1.0	5.37	4.52	5.71	4.92	4.44	4.65	4.92	4.56
5.0	26.85	22.62	28.54	24.61	22.20	23.23	24.61	22.81
10.0	53.71	45.25	57.08	49.21	44.4	46.47	49.21	45.62
25.0	134.26	113.12	142.69	123.03	111.01	116.17	123.03	114.05
50.0	268.53	226.24	285.39	246.06	222.02	232.34	246.06	228.10

NAA = α-naphthaleneacetic acid; 2,4-D = 2,4-dichlorophenoxyacetic acid; IAA = indole-3-acetic acid; IBA = indole-3-butyric acid; BAP = 6-benzylaminopurine; K = kinetin; 2iP = N-isopentenylaminopurine; Zea = zeatin.

2.3 Preparation of culture media

The most efficient way of preparing plant culture media is first to make up stock solutions of major inorganic nutrients, trace elements, iron source, vitamins, and individual plant growth regulators. An example of how to do this for the MS medium is given in *Table 6*. Store the vitamins frozen at −20 °C in small batches (e.g. enough for one litre of medium), thaw, and mix fully before use. The other stocks can be kept in a refrigerator at 4 °C, but should be frequently checked and discarded if precipitation occurs; do not store inorganic stock solutions longer than one month. It is advisable to make up small stocks of growth regulators fresh for each batch of media, as small changes in concentration due to precipitation can seriously affect growth of the cultures.

Protocol 3 outlines the preparation of media for growth of cell suspension cultures. For semi-solid media, add agar at a final concentration of 6–10 g/litre prior to autoclaving. It is important to use a good quality, bacteriological grade agar for plant cell culture work. Suitable agars for most work are New Zealand Agar (BDH Chemicals, Ltd, UK), Oxoid Bacteriological Agar No. 1 (Oxoid Ltd, UK), BiTek agar (Difco Laboratories, Detroit, MI, USA), or phytagar (GIBCO Laboratories, NY, USA). For pouring Petri dishes of agar medium, it is convenient to autoclave 500 ml batches in 1 litre conical flasks to dissolve the agar, allow the medium to cool to around 40 °C, and then pour.

Table 6. Composition and preparation of Murashige and Skoog medium

Constituent	Molarity in medium	Concentration of stock solution (mg/litre)	Volume of stock per litre of medium (ml)	Storage of stock solution
Major inorganic nutrients				
NH_4NO_3	2.06×10^{-2}	33 000		
KNO_3	1.88×10^{-2}	38 000		
$CaCl_2.2H_2O$	3.00×10^{-3}	8800	50	$+4\,°C$
$MgSO_4.7H_2O$	1.50×10^{-3}	7400		
KH_2PO_4	1.25×10^{-3}	3400		
Trace elements				
KI	5.00×10^{-6}	166		
H_3BO_3	1.00×10^{-4}	1240		
$MnSO_4.4H_2O$	9.99×10^{-5}	4460		
$ZnSO_4.7H_2O$	2.00×10^{-5}	1720	5	$+4\,°C$
$Na_2MoO_4.2H_2O$	1.00×10^{-6}	50		
$CuSO_4.5H_2O$	1.00×10^{-7}	5		
$CoCl_2.6H_2O$	1.00×10^{-7}	5		
Iron source				
$FeSO_4.7H_2O$	1.00×10^{-4}	5560	5	$+4\,°C$
$Na_2EDTA.2H_2O$	1.00×10^{-4}	7460		
Organic supplement				
myo-Inositol	4.90×10^{-4}	20 000		
Nicotinic acid	4.66×10^{-6}	100		
Pyridoxine-HCl	2.40×10^{-6}	100	5	$-20\,°C$
Thiamine-HCl	3.00×10^{-7}	100		(in 5 ml aliquots)
Glycine	3.00×10^{-5}	400		
Carbon source				
Sucrose	8.80×10^{-2}	–	Add as solid (30 g/litre)	

Protocol 3. Preparation of plant cell culture media

1. Prepare stock solutions[a] of major inorganic nutrients, trace elements, vitamins, and plant growth regulators using analytical reagent grade chemicals and double distilled water.

2. For 1 litre of liquid medium, pipette the required volumes of each stock solution into a 1 litre glass beaker on a magnetic stirrer.

3. Add sucrose as solid. Stir until it is fully dissolved, adding more water if necessary.[b]

4. Adjust the volume to ~ 950 ml with double distilled water.

5. Adjust the pH to the required value (usually 5.8–5.9) with 0.5 M NaOH.

6. Transfer to a 1 litre measuring cylinder or volumetric flask, and adjust volume to 1 litre with double distilled water.

7. Transfer back to the flask and stir for complete mixing.

8. Transfer 75 ml batches of medium to clean 250 ml conical flasks, plug with non-absorbent cotton wool or sponge plugs, cover the tops with aluminium foil, and autoclave for 15 min at 120 °C (1.06 kg/cm).

---

[a] Store inorganic stock solutions at 4 °C for a maximum of one month. Dispense vitamins in 10 ml aliquots and store at −20 °C. Prepare growth regulator stocks fresh each time; auxins are usually titrated into solution with NaOH, whereas cytokinins are dissolved in dilute NaOH or aqueous ethanol. If growth regulators are thermolabile (see *Tables 3* and *4*), add filter sterilized after the autoclaved medium has cooled.
[b] Concentrated media may be stored frozen (−20 °C) at this stage.

Note that, although the pH of most plant culture media is adjusted to between pH 5.5 and 6.0 prior to autoclaving, the final pH may be slightly lower due to the production of sugar acids during autoclaving.

The length of time required to sterilize a batch of culture medium by autoclaving will depend upon the volume of medium in the flask. For 250 ml batches, 25 minutes at 120 °C is adequate. Thermolabile growth regulators (see *Tables 3* and *4*) must be added to the medium after it has cooled to 40 °C (semi-solid media) or below.

3. Preparation and sterilization of explants

A wide-range of plant organs and tissues can be used as a source of explants for the initiation of callus cultures; this can be appreciated from a glance at *Table 2*. The choice of explant is generally dictated by the aims of the research. Some explant sources are preferable to others if embryogenic cultures are to be established for regeneration studies (see below). Likewise, for biochemical studies of secondary product synthesis, explant source may be critical. For example, alfalfa roots appear to maintain much of their secondary product metabolism even when dedifferentiated and grown as a suspension culture (76).

Protocols 4 and *5* outline sterilization procedures for seed and tuber tissues. It is often a good idea to use seedlings from sterilized seed as a source of sterile root, shoot, and leaf material. Small seed can be directly plated on callus induction medium; the seed will germinate and then the tissues will callus directly. In such cases, the callus may be of mixed origin with respect to the initiating cell type. With larger seed, it is possible to dissect sterile radicle and plumule explants directly after sterilization of the seed coat (*Protocol 6*).

Protocol 4. Sterilization of seeds for initiation of callus

1. Submerge seeds in absolute ethanol for 10 sec, or in 10% commercial bleach solution for 15 min with constant agitation.

2. Rinse in sterile distilled water (soak overnight if large seeds are to be used for dissection of radicles, see *Protocol 6*).

3. Select seeds with intact testas,[a] and submerge in 10% commercial bleach solution, containing 0.05% (v/v) Teepol detergent, for 20 min.[b]

4. Wash three times in sterile distilled water.

5. Prepare seeds for callus initiation by one of the following strategies.
    (a) For small seeds, plate out whole seeds on agar medium for callus induction or germination.
    (b) Germinate seeds on agar culture medium lacking growth regulators. Use sterile root, shoot, or leaf tissue for callus initiation.
    (c) Dissect and plate out radicle or plumule tips for callus initiation (see *Protocol 6*).

[a] Damaged testas are more readily apparent after the seed has soaked in an aqueous solution. Microbial contaminants can gain entry through damaged testas where they are subsequently protected from the bleach solution.
[b] 10% (w/v) calcium hypochlorite or 1% (w/v) bromine water are alternative sterilizing agents.

Protocol 5. Sterilization of storage organs (e.g. potato or artichoke tubers) prior to initiation of callus

1. Scrub clean under running tap-water.

2. Submerge in 10% commercial bleach solution for 30 min.

3. Wash three times in sterile distilled water.

4. Cut off an area of the skin with a sterile scalpel.

5. Remove small discs of tissue with a sterile cork borer (5–10 mm diameter) and scalpel.

6. Place discs face down on callus induction medium.

---

4. Initiation of callus and suspension cultures

4.1 Non-embryogenic cell cultures

Initiate callus by gently pressing the explant on to the surface of a suitable agar-solidified medium. Explants are usually placed horizontally, although stem sections may produce more callus if placed vertically with one cut end in the agar. Consult *Tables 1* and *2*, and Section 2.1, for details of suitable culture media. Remember that inclusion of an auxin and cytokinin will be necessary for callus growth, and that somewhat higher auxin concentrations may be required for callus initiation as compared to callus growth in some cases. After plating out explants, it is important to seal Petri plates with Clingfilm or Parafilm in order to prevent desiccation. Plates should then be incubated at approximately 25 °C either in the dark or, if beneficial, under low level illumination.

Protocol 6. Initiation of callus cultures from radicle tips of French bean (*Phaseolus vulgaris*)

1. Place seeds in 8.5 cm diameter Petri dishes (15 seeds/dish).

2. Fill dishes with 10% commercial bleach solution, ensuring seeds are fully covered. Leave for 20 min.

3. Wash the seeds in the Petri dish with three changes of 20 ml sterile distilled water.

4. Leave seeds in Petri dish to soak overnight in sterile distilled water.

5. Discard any seeds with cracked testas (use sterile forceps).

6. Re-sterilize the seeds in 10% bleach solution for 20 min.

7. Wash the seeds with three changes of 20 ml sterile distilled water.

8. Transfer the seeds to sterile Petri dishes (5 seeds/dish).

9. Using sterile forceps to hold the seed, make two cuts with a sterile scalpel in order to allow the testa to be pulled back to reveal the radicle.

10. Cut out the radicle tips (2–3 mm) and transfer to Petri dishes containing sterile distilled water (15 tips/dish).

11. Transfer the radicle tips to SH medium supplemented with 10^{-5} M pCPA, 2×10^{-6} M 2,4-D, and 10^{-7} M kinetin (5 tips/Petri dish). Seal the dishes with Parafilm and transfer to a dark cabinet at 25 °C.

If sufficient callus for subculture has not formed within three to eight weeks, it will be necessary to re-evaluate the culture medium and conditions. Subculture by removing newly formed callus with a sterile scalpel and transferring to fresh medium. Be careful not to squash the somewhat fragile callus too much when re-plating, and do not cut the callus into too many very small pieces. The optimum inoculum size varies depending on the species, but it is best to err on the large side while the culture is in the early stages of establishment. Once well established, transfer approximately 3–10 mm^3 of callus clumps from actively growing regions to the fresh medium. Most callus cultures will require regular subculture at approximately one month intervals. For newly initiated callus, it may be necessary to transfer the entire callus to the fresh medium for the first two or three subcultures. Fast growing callus cultures may require subculture at two to three week intervals.

In general, suspension cultures form readily after transfer of callus to shaken flasks of the culture medium minus agar. A large inoculum may be necessary to initiate the suspension, and agitation rates on orbital shakers should be in the range of 30–150 r.p.m. with an orbital motion stroke of 2–4 cm. If the callus culture is non-friable, very little will break off the callus clumps and the necessary inoculum size of freely suspended cells and small cell clusters may not be attained. In such cases, modifications may have to be made to the callus maintenance or induction media in order to produce a more friable callus; the approaches here are somewhat empirical.

When the newly established suspension culture has reached a suitable cell density for subculture, remove as much of the remaining unbroken callus material and large clumps as possible. This can be done by either transferring the single cells and small clumps with a sterile syringe (the orifice of which will exclude large clumps), by filtration, by allowing the large material to settle and pipetting off from the top of the culture, or simply by direct pouring. Remember, however, that there is a minimum inoculum size below which cell suspensions do not readily resume active growth following transfer, and that in some cultures most of the growth occurs on the surface of small clumps. The minimum density of cells required for cell suspension cultures depends on the rate of growth and the composition of the medium. Generally, ten per cent of initial cell density to the total volume of the culture will be sufficient. Addition of medium in which the cell line had previously been growing (conditioned medium) can sometimes stimulate growth of cell suspensions at low inoculum density. The growth rate of the culture will often increase if the cells are transferred before the culture reaches stationary phase. For some purposes, e.g. isolation of protoplasts which retain viability after electroporation (see Chapter 3), it is very important to have a vigorously growing culture.

4.2 Embryogenic cell cultures

Callus and cell cultures having the potential to produce plantlets via somatic embryogenesis are referred to as embryogenic cultures (see Chapter 5). Embryogenic cultures are widely used for obtaining transgenic plants, somaclonal variants, and mutants. Transgenic maize, wheat, rice, and oat plants have been regenerated from embryogenic callus or cell cultures transformed via biolistic bombardment-mediated DNA transfer. Protoplasts, isolated from embryogenic rice cell cultures, have been transformed via electroporation or polyethylene glycol-mediated DNA transfer for obtaining transgenic plants (see Chapter 3B).

In grasses, two different types of embryogenic calli are produced. The first type, referred to as type I embryogenic callus, is organized, compact, white or pale yellow in colour, and slow growing. The other type, referred to as type II callus, is friable, soft, somewhat translucent, and rapidly growing. Use of a dissecting microscope will make it easier to identify the non-embryonic callus from the embryogenic callus. Generally, the non-embryogenic callus is friable, fast growing, and lacking in organized structures. Embryogenic callus, on the other hand, is usually compact (except for type II callus in the case of grasses) and will show organized structures especially during the later stages of differentiation (see *Figure 1*).

Choosing the right explant is very critical for establishing embryogenic cell cultures. Immature parts of a plant (e.g. immature embryos/inflorescence, young leaves/petioles, and hypocotyls from young seedlings) have been used as explants for producing embryogenic callus, e.g. immature embryos for wheat (51), maize (55), and soybean (60); young leaves/coleoptiles for caucasian bluestem (82). *Protocol 7* describes the initiation and maintenance of embryogenic cell cultures from seed explants of the forage grass caucasian bluestem (*Bothriochloa caucasica*), and the regeneration of plants from these cultures. With some species, there is a tendency for the embyrogenic cell cultures to lose embryogenic potential (i.e. ability to regenerate plants) after a prolonged culture period. Therefore, it is essential periodically to test the regenerability of these cultures. As an alternative, cell clumps from embryogenic cultures showing good embryogenic potential can be cryo-preserved for future use (see Chapter 7).

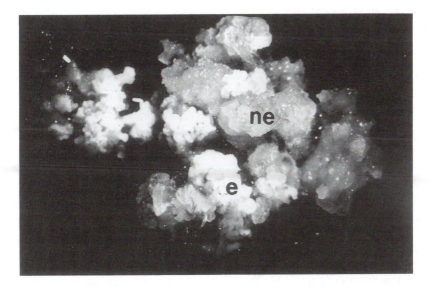

Figure 1. Embryogenic (e) and non-embryogenic (ne) callus of caucasian bluestem ($\times$ 7).

Protocol 7. Initiation and maintenance of caucasian bluestem embryogenic cell cultures

1. Remove glumes and other floral parts without damaging the seeds.
2. Wash seeds in tap-water containing a few drops of commercial detergent.
3. Rinse seeds thoroughly with tap-water.
4. Treat seeds with 70% ethanol for 30 sec to 1 min.
5. Rinse seeds thoroughly with tap-water.
6. Soak seeds in 20% commercial bleach solution containing a few drops of commercial detergent with constant agitation for 20 min.
7. Rinse seeds four times in sterile distilled water.
8. Transfer seeds to Petri dishes containing solidified CBS culture medium supplemented with 10 μM 2,4-D. CBS medium consists of MS salts (*Table 1*), B5-vitamins (*Table 1*), 2% sucrose.
9. Transfer the Petri dishes to a growth room maintained at 24 $\pm$ 2°C under low light conditions (1.2 μE.m^{-2}.s^{-1} fluorescent light), with a 16 h photoperiod.
10. After four weeks, transfer only the callus clumps produced from basal portions of leaves and coleoptiles of germinated seeds to fresh CBS medium supplemented with 10 μM 2,4-D, and incubate them under the same light and temperature conditions as described in step **9**.
11. Four weeks after the transfer (i.e. eight weeks after culture initiation) begin routine subculture of the callus to fresh medium at four week intervals.
12. After two or three subcultures, initiate cell cultures by transferring 1 g (fresh weight) of callus clumps to 15 ml of cell culture medium (CBS-L medium) contained in a 125 ml delong flask. CBS-L medium consists of CBS medium supplemented with 10 μM 2,4-D and 1 μM kinetin.
13. Incubate the flasks at 24 $\pm$ 2°C on an orbital shaker (140 r.p.m)[a] in the dark.
14. Subculture at weekly intervals.
15. To regenerate plants, transfer cell clumps to solidified CBS medium supplemented with 10 μM 2,4-D and incubate under conditions described in step **9**. After four weeks, transfer them to solidified regeneration medium (CBS without growth regulators) and incubate under conditions described in step **9**.
16. Four to eight weeks after transfer to regeneration medium, remove plantlets 10 mm or taller and transfer them to fresh regeneration medium contained in Magenta boxes (Magenta Corporation, Chicago, IL, USA) or similar containers, and grow them under higher intensity

light (50 μE.m^{-2}.s^{-1}) with a 16 h photoperiod at 24 $\pm$ 2 °C.

17. Transfer plantlets 5 cm or taller with good root systems to the green-house. Follow standard procedures (80) for greenhouse acclimatization of these plants.

[a] The size of the flask, amount of medium, and the speed set on the shaker can be important factors for obtaining fine cell cultures.

5. Measurement of growth parameters

Growth of callus and cell suspension cultures can be monitored by increase in fresh or dry weight or increase in cell number. For cell suspension cultures, increase in packed cell volume (PCV) is also a good indicator of growth.

5.1 Fresh and dry weight measurements

For callus culture, transfer the entire callus (scrape off the medium) to a pre-weighed weighing boat (or container) and determine the fresh weight. For cell suspension cultures, collect the cells on a pre-weighed nylon membrane and determine the fresh weight (determine the weight of water retained by the membrane separately and subtract this amount from the measured fresh weight). Alternatively, transfer the entire contents of the cell culture flask to a pre-weighed centrifuge tube. Spin the tube for 5–10 minutes at 200 *g*. Examine the supernatant for cells. If cells are present in the supernatant, repeat centrifugation until the supernatant is free from cells. Then carefully pipette out the entire medium without disturbing the pellet. Weigh the centrifuge tube with cells to determine the fresh weight. After measuring the fresh weight, dry the samples in an oven at 60 °C until no change in dry weight is observed.

Increase in fresh weight can also be measured without sacrificing the samples at the beginning of an experiment. Transfer callus or cells to pre-weighed Petri dishes or culture flask containing the medium. Weigh the Petri dish or the flask again and determine the weight of callus or cells added. At the end of an experiment, remove the entire callus or cells from a suspension culture and determine the final weight.

5.2 Increase in cell number

A haemocytometer can be used for determining the cell numbers in a fine suspension culture. Callus cultures and suspension cultures consisting of cell aggregates have to be macerated prior to counting the cell numbers. A variety of maceration treatments are given in ref. 77. An ideal macerating fluid for callus or cell aggregates from suspension cultures consists of equal volumes of 10% chromic acid and 10% nitric acid. Treat cell clumps for 5–30 minutes in this mixture at 60 °C. After cooling, shake the container vigorously to loosen

Table 7. Vital stains commonly used for evaluating the viability of tissue culture cells and protoplasts [a]

Compound and source	Stain preparation	Observation method	Results	Reference
Evan's blue (E-2129) [b]	0.5% in culture medium [c] or in distilled water	Examine with bright-field light microscopy	Dead cells will be stained blue, living cells will be unstained	78
Phenosafranin (P-5769) [b]	0.1% in culture medium [c] or in 0.05 M phosphate buffer at pH 5.8	Examine with bright-field light microscopy	Dead cells will be stained red, living cells will not be stained	79
Fluorescein diacetate (F-6259) [b]	Dissolve 5 mg in 1 ml of acetone—dilute this solution [d] with the culture medium [c] to a final concentration of 0.01%	Examine with fluorescence microscopy using ultraviolet light (excitation filter 450–490 nm, barrier filter 520 nm)	Fluorescence will be seen only in living cells	79

[a] For all stains, transfer a small sample of cells or protoplasts to a glass microscope slide, add one drop of the stain, place the cover glass, and wait for 5 min before microscopic observation.
[b] Numbers in parentheses indicate catalogue numbers for these compounds from Sigma Chemical Company, St. Louis, MO, USA.
[c] Preparation of these stains in the culture medium is preferred because this minimizes drastic changes in the osmolarity during staining procedure (critical for protoplasts). Such a solution, if not maintained under aseptic conditions, will be ruined due to microbial growth.
[d] Diluted solution of this stain is unstable. Therefore, it must be made fresh and used within three hours.

the cell clumps. Alternatively, add 1% (w/v) pectinase or macerozyme (Sigma Chemical Co.) to the culture medium and incubate flasks overnight on a shaker in the growth room under normal culture conditions.

5.3 Packed cell volume

Transfer the entire contents of the flask to a graduated centrifuge tube. Spin the tube at 200 g for 5–10 minutes or until supernatant is free from cells. Determine the volume of the pellet after centrifugation. Packed cell volume is expressed as a percentage of the volume of the pellet to the entire culture volume.

5.4 Cell viability

It is necessary to determine periodically the viability of cells in culture, especially if you face problems with respect to growth of the culture. Microscopic observation of cytoplasmic streaming in cells is a good indicator of cell viability, however, cell viability can also be easily and reliably evaluated using vital stains. Instructions are given in *Table 7* for the three most commonly used vital stains. For further details and methods, see Chapter 7, Section 7.

References

1. Dixon, R. A. (ed.) (1985). *Plant cell culture: a practical approach*. IRL Press, Oxford, Washington DC.
2. Vasil, I. K. (ed.) (1984). *Cell culture and somatic cell genetics of plants*, Vol. 1, Academic Press, New York.
3. Evans, D. A., Sharp, W. R., Ammirato, P. V., and Yamada, Y. (ed.) (1983). *Handbook of plant cell culture*, Vol. 1, Macmillan Publishing Co., New York.
4. Murashige, T. and Skoog, F. (1962). *Physiol. Plant.*, **15**, 473.
5. Schenk, R. U. and Hildebrandt, A. C. (1972). *Can. J. Bot.*, **50**, 199.
6. Gamborg, O., Miller, R., and Ojima, K. (1968). *Exp. Cell Res.*, **50**, 148.
7. Gresshoff, P. M. and Doy, C. H. (1974). *Planta*, **107**, 161.
8. Kao, K. N. and Michayluk, M. R. (1975). *Planta*, **126**, 105.
9. Litvay, J. D., Verma, D. C., and Johnson, M. A. (1985). *Plant Cell Rep.*, **4**, 325.
10. Nitsch, J. P. and Nitsch, C. (1969). *Science*, **163**, 85.
11. Franklin, C. I., Trieu, T. N., Gonzales, R. A., and Dixon, R. A. (1991). *Plant Cell Tiss. Organ Cult.*, **24**, 199.
12. George, E. F., Puttock, D. J. M. and George, J. H. (1987). *Plant Culture Media*, Vol. 1, Exegetics Ltd., Edington, Westbury, Wilts, England.
13. Linsmaier, E. M. and Skoog, F. (1965). *Physiol. Plant.*, **18**, 100.
14. White, P. R. (1963). *The cultivation of animal and plant cells* (2nd edn), pp. 57–63. Ronald Press, New York.
15. Dunstan, D. I. and Short, K. C. (1977). *Physiol. Plant.*, **41**, 70.
16. Negrutiu, I., Beeftink, F., and Jacobs, M. (1975). *Plant Sci. Lett.*, **5**, 293.

17. Ozias-Akins, P., Anderson, W. F., and Holbrook, C. C. (1992). *Plant Sci.*, **83**, 103.
18. Wilmar, C. and Hellendoorn, M. (1968). *Nature*, **217**, 369.
19. Shekhawat, N. S., Gordon, P. N., and Galston, A. W. (1984). *In Vitro Cell. Dev. Biol.*, **20**, 707.
20. Havel, L. and Kolar, Z. (1983). *Plant Cell Tiss. Organ Cult.*, **2**, 349.
21. Hooker, M. P. and Nabors, M. W. (1977). *Z. Pflanzenphysiol.*, **84**, 237.
22. Stafford, A., Smith, L., and Fowler, M. W. (1985). *Plant Cell Tiss. Organ Cult.*, **4**, 83.
23. Lindemann, E. G. P., Gunckel, J. E., and Davidson, O. W. (1970). *Am. Orchid Soc. Bull.*, **39**, 1002.
24. Hankes, P. A. A., Krijbolder, L., Libbenga, K. R., Wijnsma, R., Aremge, T. N., and Verpoorte, R. (1985). *Plant Cell Tiss. Organ Cult.*, **4**, 199.
25. Sondahl, M. R., Nakamura, T., Medina-Filho, H. P., Carvalho, A., Fazuoli, L. C., and Costa, W. M. (1984). In *Handbook of plant cell culture* (ed. P. V. Ammirato, D. A. Evans, W. R. Sharp, and Y. Yamada), Vol. 3, pp. 564–90. Macmillan Publishing Co., New York.
26. Paterson, K. E. and Rost, T. L. (1979). *Am. J. Bot.*, **66**, 463.
27. Aziz, H. A. and McCown, B. H. (1985). *HortScience*, **20**, 540.
28. Steward, F. C. and Mapes, M. O. (1971). *Bot. Gaz.*, **132**, 65.
29. Konar, R. N. and Singh, M. N. (1979). *Z. Pflanzenphysiol.*, **95**, 87.
30. Misawa, N., Yamano, Y., Ohyama, K., and Komano, T. (1982). In *Proc. 5th Int. Cong. Plant Tiss. Cell Cult.*, pp. 607–8. Jap. Assoc. Tiss. Cult., Tokyo.
31. Lamport, D. T. A. (1964). *Exp. Cell Res.*, **33**, 195.
32. Amberger, L. A., Palmer, R. G., and Shoemaker, R. C. (1992). *Crop Sci.*, **32**, 1103.
33. Smith, R. H. and Price, H. J. (1978). *Plant Physiol.*, **61**, 19.
34. Wessels, D. C. J., Groenewald, E. G., and Koeleman, A. (1976). *Z. Pflanzenphysiol.*, **78**, 141.
35. Gamborg, O. L. and Eveleigh, D. E. (1968). *Can. J. Biochem.*, **46**, 417.
36. Chang, W. and Chiu, P. (1978). *Z. Pflanzenphysiol.*, **89**, 91.
37. Rogers, S. M. D. and Lineberger, R. D. (1981). *HortScience*, **16**, 426.
38. Johnson, J. L. and Emino, E. R. (1979). *Hortic. Sci.*, **14**, 605.
39. Litz, R. E., Knight, R. J., and Garzit, S. (1984). *Sci. Hortic.*, **22**, 233.
40. Atanassov, A. and Brown, D. C. W. (1984). *Plant Cell Tiss. Organ Cult.*, **3**, 149.
41. Gamborg, O. L., Shyluk, J. P., Fowke, L. C., Wetter, L. R., and Evans, D. (1979). *Z. Pflanzenphysiol.*, **95**, 255.
42. Grimes, H. D. and Hodges, T. K. (1990). *J. Plant Physiol.*, **136**, 362.
43. Vasil, V. and Vasil, I. K. (1982). *Am. J. Bot.*, **69**, 1441.
44. Shneyour, Y., Zelcer, A., Izhar, S., and Beckmann, J. S. (1984). *Plant Sci. Lett.*, **33**, 293.
45. Tisserat, B. (1984). In *Handbook of plant cell culture* (ed. W. R. Sharp, D. A. Evans, P. V. Ammirato, and Y. Yamada), Vol. 2, pp. 505–45. Macmillan Publishing Co., New York.
46. Veliky, I. A. and Martin, S. M. (1970). *Can. J. Microbiol.*, **16**, 223.
47. Kshirsagar, M. K. and Metha, A. R. (1978). *Phytomorph.*, **28**, 50.
48. Ho, W. J. and Vasil, I. K. (1983). *Ann. Bot.*, **51**, 719.
49. Wernicke, W. and Brettell, R. (1980). *Nature*, **287**, 138.
50. Tsai, C. H. and Kinsellar, J. E. (1981). *Ann. Bot.*, **48**, 549.

51. Redway, F. A., Vasil, V., and Vasil, I. K. (1990). *Plant Cell Rep.*, **8**, 714.
52. Lai, P. C. and Yie, S. T. (1980). *J. Agric. Res. China*, **29**, 157.
53. MacCarthy, J. J. and Staba, E. J. (1985). *Ann. Bot.*, **56**, 205.
54. Chavez, V. M., Litz, R. E., and Norstog, K. (1992). *Plant Cell Tiss. Organ Cult.*, **30**, 99.
55. Armstrong, C. L. and Phillips, R. L. (1988). *Crop Sci.*, **28**, 363.
56. Trigiano, R. N., May, R. A., and Conger, B. V. (1992). *In Vitro Cell. Dev. Biol.*, **28**, 187.
57. Shetty, K. and Asano, Y. (1991). *J. Plant Physiol.*, **139**, 82.
58. Trigiano, R. N. and Conger, B. V. (1987). *J. Plant Physiol.*, **130**, 49.
59. Armstrong, C. L. and Green, C. E. (1985). *Planta*, **164**, 207.
60. Ranch, J. P., Oglesby, L., and Zielinski, A. C. (1985). *In Vitro Cell. Dev. Biol.*, **21**, 653.
61. Lam, S. L. (1977). *Am. Potato J.*, **54**, 575.
62. Protacio, C. M. and Flores, H. E. (1992). *In Vitro Cell. Dev. Biol.*, **28**, 81.
63. Mohamed, M. F., Read, P. E., and Coyne, D. P. (1992). *J. Am. Soc. Hortic. Sci.*, **117**, 668.
64. Pfister, J. M. and Widholm, J. M. (1984). *Hortic. Sci.*, **19**, 852.
65. Slimmon, T., Qureshi, J. A., and Saxena, P. K. (1991). *Plant Cell Rep.*, **10**, 587.
66. Hagen, S. R., Muneta, P., Augustin, J., and LeTourneau, D. (1991). *Plant Cell Tiss. Organ Cult.*, **25**, 45.
67. Ellis, D. D., Barczynska, H., McCown, B. H., and Nelson, N. (1991). *Plant Cell Tiss. Organ Cult.*, **27**, 281.
68. Nabors, M. W., Heyser, J. W., Dykes, T. A., and DeMott, K. J. (1983). *Planta*, **157**, 385.
69. Kohlenbach, H. W., Wenzel, G., and Hoffmann, F. (1982). *Z. Pflanzenphysiol.*, **105**, 131.
70. Binns, A. N., Chen, R. H., Wood, H. N., and Lynn, D. G. (1987). *Proc. Natl. Acad. Sci. USA*, **84**, 980.
71. Lynn, D. G., Chen, R. H., Manning, K. S., and Wood, H. N. (1987). *Proc. Natl. Acad. Sci. USA*, **84**, 615.
72. Stonier, T., Macgladrie, K., and Shaw, G. (1979). *Plant Cell Environ.*, **2**, 79.
73. Parthier, B. (1990). *J. Plant Growth Reg.*, **9**, 57.
74. Jacobs, M. and Rubery, P. H. (1988). *Science*, **241**, 346.
75. Gross, D. (1975). *Phytochemistry*, **14**, 2105.
76. Kessmann, H., Choudhary, A. D., and Dixon, R. A. (1990). *Plant Cell Rep.*, **9**, 38.
77. Berlin, G. P. and Miksche, J. P. (1976). *Botanical microtechnique and cytochemistry*, pp. 128–9. The Iowa State University Press, Ames, Iowa.
78. Taylor, J. A. and West, D. W. (1980). *J. Exp. Bot.*, **31**, 571.
79. Widholm, J. M. (1972). *Stain Technol.*, **47**, 189.
80. Pierik, R. L. M. (1987). *In vitro culture of higher plants*, pp 127–32. Martinus Nijhoff Publishers, Dordrecht, Boston, Lancaster.
81. Han, K. and Stephens, L. C. (1992). *Plant Cell Tiss. Organ Cult.*, **31**, 211.
82. Franklin, C. I., Trieu, T. N., and Gonzales, R. A. (1990). *Plant Cell Rep.*, **9**, 443.

2

Isolation, culture, and regeneration of protoplasts

N. W. BLACKHALL, M. R. DAVEY, and J. B. POWER

1. Introduction

Protoplasts provide the starting point for many of the techniques of genetic manipulation of plants, in particular the induction of somaclonal (protoclonal) variation, somatic hybridization, and transformation. Such experiments can be considered to consist essentially of three stages; protoplast isolation, the genetic manipulation event involving protoplast fusion or gene uptake and, finally, protoplast culture and regeneration of fertile plants.

Theoretically, all plant cells contain the genetic information necessary for their development into whole plants. However, it is not always possible for this 'totipotency' to be expressed, giving rise to the concept of 'morphogenic competence'. In general, three principal factors govern competence, these being the plant genotype, ontogenetic state of the explant source, and the cultural environment. The latter includes medium composition and growth conditions.

Protoplasts have been isolated by mechanical disruption or by enzymatic degradation of their surrounding cell walls, enzymatic degradation being the most general procedure. Although almost any explant of most plant species can be used as a source of protoplasts, the ability to isolate protoplasts capable of sustained division and plant regeneration is still restricted to a limited number of species/explant combinations (1). Leaf tissues are frequently employed as source materials for protoplast isolation. However, sustained division leading to plant regeneration is not routine for mesophyll-derived protoplasts of monocotyledonous species, with the exception of an example in rice (2). Consequently, embryogenic cell suspension cultures provide the most commonly used supply of competent cells for cereals and grasses. It is frequently observed, when isolating protoplasts directly from leaves, that young tissues release protoplasts with the highest viability. In this respect, axenic shoot cultures are being used increasingly as an alternative to glasshouse grown plants, since it is easier to regulate their growth conditions.

Furthermore, axenic shoots provide a continuous supply of juvenile tissues, which is particularly important in the case of woody species (3).

2. Protoplast isolation

2.1 Enzyme treatment

The optimum conditions and enzyme treatments for a particular plant genotype/explant combination have been determined empirically in most cases (see *Table 1* for details). Usually, protoplast isolations are performed at a temperature of 25–28 °C for a period of either 2–6 hours (short duration) or 12–20 hours (overnight). A short plasmolysis treatment, typically involving a one hour incubation in CPW salts solution (4) containing the same osmoticum as the enzyme mixture, but lacking enzymes, has been found to be beneficial in maintaining protoplast viability and reducing spontaneous fusion in several species, especially when the protoplasts are isolated directly from explants.

2.2 Protoplast purification

Following enzyme treatment, a mixture is obtained which is composed of released protoplasts, undigested cells, and cellular debris suspended in the enzyme solution. Several purification procedures have been developed, the most successful of which consists of passing the digested material through nylon or metal sieves of decreasing pore size to remove any undigested cell clumps. This is followed by centrifugation and resuspension of the protoplasts in a washing medium containing the same concentration of osmoticum as the enzyme mixture, to remove cellular debris. Centrifugation and resuspension

Table 1. Enzyme mixtures and conditions for protoplast isolation

Species	Tissue	Enzyme mixture	Comments
Arabidopsis thaliana	Leaves	1% (w/v) Cellulase R10, 0.25% (w/v) Macerozyme R10, 8 mM $CaCl_2$, 0.4 M mannitol, pH 5.5	Cut leaves into 1 mm^2 pieces, plasmolyse for 60 min in 0.5 M mannitol; digest for 14–18 h, 25 °C, dark
Hordeum vulgare	Embryogenic cell suspension	2% (w/v) Cellulase Onozuka RS, 0.05% (w/v) Pectolyase Y23, 976 mg/litre MES, CPW13M (4), pH 5.6	16 h
Brassica napus	Hypocotyl, stem, and petiole	1% (w/v) Cellulysin, 0.1% (w/v) Macerozyme R10, in K3 solution (22) containing 136.9 g/litre sucrose, pH 5.6	See *Protocol 2*

Table 1. *Continued*

Species	Tissue	Enzyme mixture	Comments
Dimorphotheca aurantiaca	Leaves, seedling cotyledons	2% (w/v) Meicelase, 2% (w/v) Rhozyme HP 150, 0.03% (w/v) Macerozyme R10, CPW8M, antibiotics,[a] pH 5.8	Plasmolyse for 1 h, 5–6 h static incubation for leaves, 10–12 h for cotyledons
Dimorphotheca aurantiaca	Ray florets, callus, stem	0.8% (w/v) Cellulase R10, 0.4% (w/v) Driselase, 0.8% (w/v) Rhozyme HP 150, 0.4% (w/v) Macerozyme R10, CPW13M, antibiotics,[a] pH 5.8	Plasmolyse for 1 h, 10 h static incubation for ray florets; 4–5 h shaking[b] followed by 8–10 h static incubation for stems and cotyledons
Glycine max	Hypocotyls from 5- to 7-day-old seedlings	As for *Dimorphotheca* leaves, but dissolved in CPW13M solution	Plasmolyse 30 min, incubate with enzyme for 16 h with shaking[b] (10 r.p.m.) in the light (2.5 μE.m.$^{-2}$.s^{-1})
Glycine canescens	Cotyledons from 10-day-old seedlings	1% (w/v) Rhozyme HP150, 2% (w/v) Cellulase R10, 0.1% (w/v) Pectolyase Y23, CPW9M, pH 5.6	Plasmolyse for 20 min, incubate for 16 h in the dark
Nicotiana tabacum cv Xanthi nc	Expanded leaves	1.5% (w/v) Meicelase, 0.05% (w/v) Macerozyme R10, CPW13M, antibiotics,[a] pH 5.8	Lower epidermis removed by peeling; 16 h static incubation
Oryza sativa v. Taipei 309	Embryogenic cell suspension	0.33% (w/v) Cellulase RS, 0.033% (w/v) Pectolyase Y23, 325.4 mg/litre MES, CPW13M, pH 5.6	See *Protocol 1*
Rudbeckia hirta	Leaves cotyledons, ray florets	0.8% (w/v) Cellulase R10, 0.4% (w/v) Driselase, 0.8% (w/v) Rhozyme HP 150, 0.2% (w/v) Macerozyme R10, CPW8M, antibiotics,[a] (CPW13M for ray florets), pH 5.8	Plasmolyse for 1 h, then 3–4 h enzyme incubation with shaking,[b] followed by 8–10 h stationary incubation in the dark (23 °C)

[a] Antibiotic mixture: 400 mg/litre ampicillin, 10 mg/litre gentamycin, 10 mg/litre tetracyclin.
[b] Shaker: 30 r.p.m., 2 cm throw.

are repeated until a purified protoplast preparation is obtained. A flotation step may be employed for protoplast systems containing excessive amounts of debris, especially leaf mesophyll-derived protoplasts. In this case, the protoplasts, after sieving and a single wash to remove the enzyme, are gently mixed with 21% (w/v) sucrose in CPW salts solution (CPW21S). On centrifugation (100 *g*; 10 min), protoplasts collect as a band at the top of the sucrose solution, whereas organelles and cellular debris pellet or remain suspended. Centrifugation speed and time may need to be adjusted for different protoplast systems.

2.3 Visualization of cell walls

A rapid assessment of the efficiency of cell wall removal can be obtained by observing protoplasts under the light microscope. Spherical shape and the absence of birefringence usually indicate complete removal of the cell wall. Additionally, any remaining cell wall material can be stained with Calcofluor White (5) or the fluorescent brightener Tinapol (6). Mix one drop of a 0.1% (w/v) solution of the stain in CPW salts with 13% (w/v) mannitol (CPW13M) with an equal volume of the protoplast suspension on a microscope slide. Examine the protoplasts under blue illumination, using a fluorescence microscope, after five minutes incubation at 22 °C. Cell walls stained with Tinapol fluoresce yellow; those stained with Calcofluor White produce an intense blue fluorescence when examined under UV illumination.

2.4 Determination of protoplast viability

Fluorescein diacetate (FDA; 7) is often used to determine protoplast viability. FDA molecules are non-fluorescent, but are able to pass freely across the plasma membrane. Esterases within the cytoplasm of living cells cleave FDA to release the fluorescent compound fluorescein, which is unable to pass out through the plasma membrane of viable protoplasts. Mix 1 ml of a 5 mg/ml stock solution of FDA with 10 ml CPW salts solution to prepare a working dilution. Mix equal volumes of the working dilution of FDA and the protoplast suspension. After five minutes at 22 °C, examine the protoplasts with a fluorescence microscope, using blue illumination. Viable protoplasts emit a green/yellow fluorescence, whereas the non-viable protoplasts remain unstained. See Chapters 1 and 7 for other methods of determining viability.

2.5 Counting protoplasts

Count plant protoplasts using a modified Rosenthal haemocytometer. Prepare the haemocytometer by moistening the sides of the chamber and then sliding on the special cover-slip while pressing down towards the body of the chamber. Correct separation between counting area and cover-slip is obtained when a diffraction pattern (Newton's rings) is observed at the point of contact of the cover-slip with the body of the haemocytometer. Resuspend

---

Protocol 1. Isolation of protoplasts from embryogenic cell suspension cultures of rice (*Oryza sativa* v. Taipei 309)

Equipment and reagents

- Exponentially growing cell suspension cultures three to seven days after subculture, initiated and maintained as described (8)
- Autoclaved sieves constructed from stainless steel, nylon, or polyester fabric with pore sizes of 500, 64, 45, and 30 μm
- CPW salts solution with 130 g/litre mannitol (CPW13M)
- Enzyme mixture (see *Table 1*)
- 14 cm Petri dishes
- Sealing film, e.g. Nescofilm or Parafilm M
- Sterile Pasteur pipettes and volumetric pipettes
- Orbital platform shaker (30 r.p.m., 2 cm throw) with temperature control (27 °C)

Method

1. Filter the cell suspension through a sieve of 500 μm pore size into a pre-weighed 14 cm Petri dish. Remove the medium with a sterile 10 ml pipette to leave the cells in the dish.

2. Re-weigh the Petri dish and add the enzyme mixture (10 ml/g fresh weight of cells).

3. Seal the Petri dish with Nescofilm or Parafilm M and incubate overnight (15 h) in the dark at 27 °C on an orbital shaker (30 r.p.m.), followed by stationary incubation for 1 h under the same conditions.

4. Sequentially filter the cell suspension/enzyme mixture through sieves of 64, 45, and 30 μm mesh to remove undigested cells.

5. Collect the protoplasts in the filtrate, wash three times by centrifugation (80 *g*, 10 min), and resuspend in CPW13M solution.

Protocol 2. Protoplast isolation from hypocotyls of *Brassica napus* L. var *oleifera*

Equipment and reagents

- Seeds of *Brassica napus* L. var. *oleifera* spring cultivars Diplom, Line, or Optima
- 1.5% (v/v) calcium hypochlorite solution
- Sterile distilled water
- Germination medium; MS based medium (9) supplemented with 10 g/litre sucrose
- Plasmolysis solution; 300 mM sorbitol, 50 mM CaCl₂
- Enzyme mixture (see *Table 1*)
- Salts solution (10) containing 160 g/litre sucrose
- W5 solution (11)

Method

1. Surface sterilize seeds by immersion in calcium hypochlorite solution for 20 min with continuous shaking (30 r.p.m.) at 22 °C.

Protocol 2. *Continued*

2. Wash the seeds three times by rinsing with sterile distilled water.

3. Germinate seeds in 9 cm Petri dishes containing germination medium for four to five days in the dark at 25 °C.

4. Excise 20–30 mm long hypocotyls and cut transversely into segments 0.5–1.0 mm in length. Place the segments inside a sieve of 60–100 μm pore size and immerse in plasmolysis solution for 1 h at 22 °C.

5. Transfer the segments to enzyme solution (10 ml enzyme/g fresh weight of tissue). Incubate for 16 h at 22 °C in the dark on a shaker (30 r.p.m.)

6. Filter the protoplasts through a sieve of 50 μm pore size and dispense as 12 ml aliquots. Add 4 ml salts solution and centrifuge (100 *g*; 7 min). With-draw the floating protoplasts with a Pasteur pipette, dilute with W5 solution, and centrifuge (75 *g*; 5 min).

the protoplasts in a known volume (usually 10 ml) of washing solution (e.g. CPW13M). Introduce a sample beneath the cover-slip, to fill the counting area. Care needs to be taken not to overfill the chamber. Examination of the chamber using a microscope will reveal a grid of small squares with a triple line every fourth line. The triple-lined square encloses 16 smaller squares. The number of protoplasts enclosed by the triple-lined square is counted (n), including those touching the top and left edges, but not the bottom or right edges. Calculate the number of protoplasts per millilitre as $5n \times 10^3$, the total yield for a volume of 10 ml being $5n \times 10^4$.

3. Protoplast culture

3.1 Media

The nutritional requirements of protoplasts and cell suspension cultures are often similar. Consequently, protoplast culture media are frequently variants of the media used for cell culture, namely those based on the MS (9), B5 (12), and KM (13) formulations (see Chapter 1). In order to induce sustained protoplast division, some media modifications may be necessary for certain species and cultivars, such as modifications to KM medium for the growth of protoplasts of *Medicago sativa* (14). Ammonium ions are particularly detrimental to protoplast survival and have been reduced or removed from many media. Similarly, other workers have found it necessary to change the microelements and organic components of published formulations.

The osmotic pressure of the enzyme mixture, protoplast washing solutions, and culture media is adjusted by the addition of sugars or sugar alcohols. For isolation, it is common to adjust the osmotic pressure by addition of mannitol up to 130 g/litre. Slightly higher osmotic pressures can be obtained, if required, by using sorbitol, with the advantage of less crystallization of the sugar alcohol from saturated solutions. A carbon source must be added to the culture medium, since protoplast-derived tissues are generally non-photosynthetic. Sucrose is most commonly used for this purpose. Glucose is frequently employed as both osmoticum and carbon source; cultured proto-plasts synthesizing new cell walls rapidly remove this sugar from the medium. For some species, this reduction in the osmotic pressure of the medium is beneficial, whereas for others it is necessary to either supplement or to replace the glucose with sucrose or with a non-metabolizable sugar alcohol such as mannitol. For rice protoplasts, there are reports (15) that maltose as the carbon source may lead to a higher frequency of plant regeneration compared to sucrose.

3.2 Culture procedures

In general, protoplasts of most species prefer to be embedded at a density between 5.0×10^2 and 1.0×10^6/ml in a medium made semi-solid with agar or agarose, rather than being cultured in liquid medium. This is presumably because of the enhanced support provided by agar or agarose, which encour-ages cell wall development. However, liquid media allow faster diffusion of nutrients and waste products, as well as facilitating reduction of the osmotic pressure as the protoplasts grow and resynthesize cell walls. The purity of the agar or agarose is important, with low gelling temperature agaroses such as SeaPlaque (FMC BioProducts, Rockland, ME, USA) or Sigma types VII and IX being used extensively. Several protocols have been devised to exploit both liquid and semi-solid systems. The most frequently used techniques are as follows:

(a) Embedding in agar or agarose. Protoplasts are suspended at the required plating density in culture medium containing molten (40 °C) agarose (1.2% w/v). The suspension is dispensed into Petri dishes (3 or 5 cm diameter) and allowed to cool and solidify. The dishes are sealed with Nescofilm or Parafilm and incubated in the culture room, either in the dark or under low intensity illumination (e.g. 7 μE.m^{-2}.s^{-1} from daylight fluorescent tubes) with a suitable photoperiod. The agar or agarose layer containing the embedded protoplasts may be cut into sections and the latter transferred to larger Petri dishes (e.g. 9 cm) containing liquid cul-ture medium of the same composition. Alternatively, the molten agarose medium containing the suspended protoplasts is dispensed as droplets (50–150 μl) in the bottom of Petri dishes. After solidification, the droplets are bathed in liquid medium of the same composition.

(b) Liquid-over-agar or agarose. A thin layer of agarose solidified medium is formed in the bottom of a Petri dish and liquid medium containing the protoplasts at twice the required plating density is poured over the agarose layer.

(c) Hanging drop culture. This permits culture of small numbers of protoplasts, often at high density, in liquid medium. Small droplets, 20–40 μl
in volume, containing the protoplasts are dispensed into the lids of Petri
dishes. Culture medium or sterile distilled water is placed in the base of
the Petri dish. The lid with the droplets of protoplasts is inverted and
placed over the base before the Petri dish is sealed with Nescofilm.

Alginate is an alternative gelling agent for use with protoplasts which are
particularly heat sensitive, e.g. those of *Arabidopsis thaliana* (16). Alginate
may also be used if it is necessary to depolymerize the medium to release
developing protoplast-derived colonies. Solutions containing alginate solidify
in the presence of Ca^{2+} ions. Frequently, a high Ca^{2+} concentration is used to
solidify the alginate-containing medium; the protoplasts are then maintained
in a medium with a lower Ca^{2+} level which is just sufficient to keep the
alginate solidified. The alginate medium may be solidified either as a thin
layer by pouring over an agar layer containing Ca^{2+} ions, or as beads, by
dropping into liquid medium containing the Ca^{2+} ions. When depolymer-
ization is required, the Ca^{2+} ions are removed by a brief exposure to sodium
citrate and the released cell colonies are washed free of alginate and citrate.

A minimum plating density is essential for protoplast division and sustained
growth, a density of greater than 1.0×10^5 protoplasts/ml often being required. Nurse cell techniques have been used to promote protoplast division
at low plating densities. The nurse techniques also facilitate the culture of
small protoplast populations. There are two types of nurse culture:

(a) Protoplasts or cells capable of rapid division, from the same genus,
species, or variety can be used as a nurse culture. Protoplasts or cells from
embryogenic cell suspensions are generally preferable to those from non-
embryogenic sources. It is essential to separate physically the two protoplast populations, either with a filter membrane (5–12 μm pore size) or
nylon or polyester sieve material. A simple culture chamber has been
described (17) in which the protoplasts of interest are cultured in a 30 $\times$
7.5 mm membrane cylinder. The latter is held in a vertical position in the
centre of a 5 cm diameter Petri dish by embedding in an agarose medium.
The nurse protoplasts are plated outside the cylinder in a liquid medium
over the agarose layer. Commercially manufactured Millicell-CM units
(Millipore, Bedford, MA, USA) are also suitable for culturing protoplasts in the presence of nurse protoplasts or cells, since they are transparent and so permit examination of the developing protoplasts/cell
colonies. Alternatively, membranes of 0.22 μm pore size can be used to
separate physically the two populations (18).

(b) X- or γ-irradiated protoplasts can be used as a nurse. Such protoplasts are
incapable of sustained growth and division and, if necessary, can be
cultured directly with the test protoplasts (19). Alternatively, the two
populations can be separated physically (20).

In addition to utilizing nutrients from the culture medium, dividing protoplasts also release growth promoting factors, particularly amino acids, into the surrounding environment. Such compounds probably contribute to the 'nurse' effect. 'Conditioned' medium can be prepared by culturing protoplasts in liquid medium. Subsequently, the protoplasts are removed, the medium filter sterilized, and reused to culture the test protoplasts. The period for optimum conditioning may need to be determined empirically, with the result that the medium should be harvested at 24 hour intervals from the beginning of the culture period.

Protocol 3. Culture and plant regeneration from protoplasts isolated from embryogenic cell suspension cultures of rice (*Oryza sativa* v. Taipei 309)

1. Isolate protoplasts as described in *Protocol 1*.

2. Resuspend the protoplasts in KPR medium (K8P medium (14) supplemented with an extra 300 μg/litre 2,4-dichlorophenoxyacetic acid [2,4-D]) at a density of 5.0×10^5/ml in a centrifuge tube. Heat shock by incubating in a water-bath at 45 °C for 5 min followed by 30 sec in ice.

3. Wash the protoplasts twice by centrifugation at 80 *g*, followed by resuspension in fresh KPR medium.

4. Resuspend the protoplasts at a density of 3.5×10^5/ml in KPR medium made semi-solid with 12 g/litre SeaPlaque agarose (mix equal volumes of a 24 g/litre agarose solution with double strength KPR medium).

5. Culture 2 ml of protoplast suspension in 3.5 cm Petri dishes, sealed with Nescofilm, in the dark at 27 °C.

6. After 14 days, divide the agarose layers from each dish into four portions and transfer each portion to a separate 5 cm Petri dish containing 3 ml liquid KPR medium. Incubate in the dark at 27 °C until cell colonies 0.5–10 mm in diameter have developed.

7. Transfer the protoplast-derived colonies to 5×5 square well plastic Petri dishes (Sterilin, Hounslow, UK, four colonies per well). The wells should each contain 2 ml MSKN medium (MS medium supplemented with 2.0 mg/litre α-naphthaleneacetic acid (NAA), 500 μg/litre zeatin, and 30 g/litre sucrose, solidified by the addition of 12 g/litre SeaPlaque agarose). Incubate at 27 °C in the dark.

8. 7–14 days later, transfer shoots with roots and coleoptiles to MS medium supplemented with 30 g/litre sucrose and 4 g/litre Sigma type I agarose. Incubate at 25 °C in the light (5 μE.m^{-2}.s^{-1}, 16 h photoperiod, daylight fluorescent tubes).

Protocol 3. *Continued*

9. Transfer rooted plants to 8 cm pots containing a 12:1 (v/v) mixture of Levington M3 soil-less compost (Fisons PLC, Ipswich, UK) and Silvaperl Perlite (Silvaperl Ltd, Gainsborough, UK) and grow to maturity in a glasshouse (natural daylight supplemented with 30 $\mu E.m^{-2}.s^{-1}$ of daylight fluorescent illumination; day and night temperature maxima of 30 °C and 18 °C respectively).

Protocol 4. Plant regeneration from mesophyll protoplasts isolated from axenic shoot cultures of apple (*Malus X domestica* Borkh) rootstock M9 (21)

1. Prepare quadruple strength SeaPlaque agarose by dissolving in distilled water at a concentration of 25 g/litre; sterilize by autoclaving.

2. Prepare and filter sterilize a 100-fold concentrated organic additive stock solution consisting of: 200 mg/litre glycine, 200 mg/litre thiamine-HCl, 100 mg/litre pyridoxine-HCl, 100 mg/litre nicotinic acid, 50 mg/litre riboflavin, 5 mg/litre D-biotin, 1 mg/litre cyanocobalamin (Vitamin B_{12}), 10 mg/litre folic acid, 100 mg/litre D-pantothenic acid (hemicalcium salt), 5 g/litre casein enzymatic hydrolysate (from bovine milk), and 50 g/litre *myo*-inositol in distilled water.

3. Prepare liquid MS salts-based culture medium (9; see Chapter 1) supplemented with 2.0 mg/litre NAA, 250 μg/litre 6-benzylaminopurine (BAP), 1% (v/v) of the organic additive stock, and 90 g/litre mannitol, pH 5.8. Also prepare the same culture medium to double strength.

4. When required, mix 10 ml double strength liquid culture medium with 10 ml liquefied (molten) quadruple strength agarose to produce 2 × agarose plating medium. Maintain at a temperature of not less than 45 °C to prevent gelling.

5. Isolate M9 mesophyll protoplasts according to (21). Resuspend the protoplasts in CPW13M solution and purify by layering on to CPW21S solution prior to centrifuging (100 *g*; 10 min). Collect the protoplasts from the CPW13M/CPW21S interface and resuspend in single strength liquid MS salts-based culture medium at a density of 1.0 × 10^6 protoplasts/ml. Mix the protoplast suspension with an equal volume of the 2 × agarose plating medium, which has been allowed to cool to approximately 30 °C, and dispense as 5 ml layers in 3.5 cm diameter Petri dishes. Seal the dishes with Nescofilm or Parafilm and incubate at 25 °C in the dark.

6. After 14 days, cut each agarose layer, containing dividing protoplasts, into four and transfer to 5.0 cm Petri dishes containing liquid culture

medium supplemented with mannitol at a concentration of 67.5 g/litre (the agarose culture medium contained mannitol at 90 g/litre).

7. After a further seven days, reduce the osmotic pressure by replacing the medium with fresh liquid medium containing 51 g/litre mannitol.

8. 60 days after protoplast isolation, transfer the protoplast-derived tissues to MS medium supplemented with 2.0 mg/litre NAA, 0.5 mg/litre BAP, and solidified with 8 g/litre agar.

9. Transfer, at 12 weeks from the date of protoplast isolation, the largest tissues to shoot regeneration medium. The latter is MS salts-based medium supplemented with 0.2 mg/litre thiamine-HCl, 1.0 mg/litre pyridoxine-HCl, 1.0 mg/litre nicotinic acid, 500 mg/litre casein enzymatic hydrolysate, 126 mg/litre phoroglucinol, 10 µg/litre NAA, 2 mg/litre BAP, solidified with 8 g/litre agar, pH 5.8. Maintain at 25°C, under 5 µE.m^{-2}.s^{-1} of continuous daylight fluorescent illumination. Transfer to fresh medium every 21 days.

10. As shoot buds develop, detach and transfer to MS salts-based multiplication medium with 2.0 mg/litre BAP, 50 g/litre sucrose, and solidified with 1.5 g/litre Phytagel (Sigma). Subculture the shoots, when 2 cm in height, for root initiation, to half strength MS medium with 3.0 mg/litre indole-3-butyric acid (IBA) and 1.5 g/litre Phytagel. After seven days, subculture the plants to half strength hormone-free MS medium, solidified with 1.5 mg/litre Phytagel. Culture for three weeks to allow root elongation.

11. Transfer rooted plants to pots containing Levington M3 soil-less compost in a glasshouse.

Protocol 5. Culture and plant regeneration from hypocotyl protoplasts of *Brassica napus* L. var *oleifera*

1. Resuspend protoplasts obtained as described in *Protocol 2* at a density of 2.5 × 10^4/ml in K8P medium (13) containing 72.06 g/litre glucose, 496 µg/litre BAP, 93 µg/litre NAA, and 995 µg/litre 2,4-D. Culture the protoplasts in 1 ml aliquots in 5 cm Petri dishes and maintain at 25°C in the dark.

2. After three to five days, when cell wall regeneration has occurred and the first signs of cell division are observable, add 3 ml of K8P medium without 2,4-D.

3. When microcalli are visible, 14 days after protoplast isolation, transfer the microcalli to 9 cm Petri dishes containing 30 ml K3 medium (22) supplemented with 34.2 g/litre sucrose, 496 µg/litre BAP, 93 µg/litre NAA, and 243 µg/litre 2,4-D. Incubate the Petri dishes at 25°C under fluorescent light, 10 µE.m^{-2}.s^{-1}, 16 h photoperiod.

Protocol 5. *Continued*

4. After three weeks, transfer calli 2–4 mm in diameter to K3 medium (22) supplemented with 10.3 g/litre sucrose, 1.096 mg/litre zeatin, 1.126 mg/litre BAP, and 1.051 mg/litre IAA, and made semi-solid with 4 g/litre agarose (Sigma type I).

5. Root regenerating shoots by transferring to MS salts-based medium with 30 g/litre sucrose and 372.4 µg/litre NAA.

6. Grow plantlets in 8 cm diameter pots containing a 12:1 (v/v) mixture of Levington M3 soil-less compost and Silvaperl Perlite under high humidity conditions, before transfer to the glasshouse.

4. Concluding remarks

The establishment of reproducible protoplast-to-plant systems for a range of plants has allowed the pursuit of genetic manipulation experiments in a number of laboratories. The needs of the world's population for ever increasing supplies of food place immense pressure on plant biotechnology to develop new varieties with higher yields or better disease and stress tolerance. Consequently, much effort is currently being directed towards the major crop species. Efficient protoplast-to-plant systems provide the crucial baseline for the effective introduction of foreign DNA and genes into higher plants of agronomic value. Additionally, protoplasts are able to provide experimental systems for a wide-range of biochemical and molecular studies, ranging from investigations into the growth properties of individual cells, to membrane transport. The use of protoplast technology has also been extended to include bacterial and fungal systems (23).

References

1. Ochatt, S. J. and Power, J. B. (1992). In *Plant biotechnology: comprehensive biotechnology second supplement* (ed. M. W. Fowler and G. S. Warren), pp. 99–128. Pergamon Press, Oxford.
2. Gupta, H. S. and Pattanayak, A. (1993). *Bio/Technology*, **11**, 90.
3. Manders, G., Davey, M. R., and Power, J. B. (1992). *J. Exp. Bot.*, **43**, 1181.
4. Frearson, E. M., Power, J. B., and Cocking, E. C. (1973). *Dev. Biol.*, **33**, 130.
5. Galbraith, D. W. (1981). *Physiol. Plant.*, **53**, 111.
6. Cocking, E. C. (1985). *Bio/Technology*, **3**, 1104.
7. Widholm, J. M. (1972). *Stain Technol.*, **47**, 189.
8. Finch, R. P., Lynch, P. T., Jotham, J. P., and Cocking, E. C. (1991). In *Biotechnology in agriculture and forestry* (ed. Y. P. S. Bajaj), Vol. 14, pp. 251–68. Springer-Verlag, Berlin.
9. Murashige, T. and Skoog, F. (1962). *Physiol. Plant.*, **15**, 473.
10. Banks, M. S. and Evans, P. K. (1976). *Plant Sci. Lett.*, **7**, 409.

11. Menczel, L., Nagy, I., Kiss, Z. R., and Maliga, P. (1981). *Theor. Appl. Genet.*, **59**, 191.

12. Gamborg, O. L., Miller, R. A., and Ojima, K. (1968). *Exp. Cell Res.*, **50**, 151.

13. Kao, K. N. and Michayluk, M. (1975). *Planta*, **126**, 105.

14. Gilmour, D. M., Golds, T. J., and Davey, M. R. (1989). In *Biotechnology in agriculture and forestry*, Vol. 8, *Plant protoplasts and genetic engineering I.* (ed. Y. P. S. Bajaj), pp. 370–88. Springer-Verlag, Berlin.

15. Biswas, G. C. G. and Zapata, F. J. (1992). *J. Plant Physiol.*, **139**, 523.

16. Damm, B., Schmidt, R., and Willmitzer, L. (1989). *Mol. Gen. Genet.*, **217**, 6.

17. Gilmour, D. M., Davey, M. R., Cocking, E. C., and Pental, D. (1987). *J. Plant. Physiol.*, **126**, 457.

18. Guiderdoni, E. and Chair, H. (1992). *Plant Cell Rep.*, **11**, 618.

19. Raveh, D., Huberman, E., and Galun, E. (1973). *In Vitro Cell. Dev. Biol.*, **9**, 216.

20. Cella, R. and Galun, E. (1980). *Plant Sci. Lett.*, **19**, 243.

21. Patat-Ochatt, E. M., Ochatt, S. J., and Power, J. B. (1988). *J. Plant Physiol.*, **133**, 460.

22. Nagy, J. J. and Maliga, P. (1976). *Z. Pflanzenphysiol.*, **78**, 453.

23. Peberdy, J. F. (1989). *Mycological Res.*, **93**, 1.

3

Applications of protoplast technology

3A. Fusion and selection of somatic hybrids

N. W. BLACKHALL, M. R. DAVEY, and J. B. POWER

1. Introduction

Somatic hybridization of plants involves four discrete stages; protoplast isolation, protoplast fusion, the regeneration of plants from selected tissues, and analysis of regenerated plants. As virtually any combination of protoplasts can be induced to undergo fusion, somatic hybridization provides a means to circumvent sexual barriers to plant breeding. It not only provides a method for generating hybrids between sexually incompatible plants, but also facilitates the genetic modification of vegetatively propagated crops, sterile or subfertile species, and plants with naturally long life cycles.

Fusion treatment results in the production of heterokaryons and homokaryons, while some protoplasts remain unfused. Heterokaryons are the fusion products relevant to plant genetic manipulation. They contain the nuclei of the two genera, species, or varieties, initially in a mixed cytoplasm. Heterokaryons may develop into hybrid cells. Like unfused plant cells, somatic hybrid cells are totipotent and are therefore capable of developing, via embryogenesis or organogenesis, into plants. Complex nucleo–cytoplasmic combinations may follow protoplast fusion. Incompatibilities between the two nuclear genomes or the nucleo–cytoplasmic combinations may become apparent, leading to either chromosome elimination or, in extreme cases, failure of heterokaryons to undergo sustained growth and division. Mitochondrial DNA recombination often occurs to generate 'new' organelles. Chloroplasts usually segregate, with those of one partner becoming dominant; very rarely, chloroplast DNA recombinations may take place. Irradiation of one of the protoplast partners prior to fusion may enhance chromosome elimination, resulting in fusion products which retain the nuclear genome of one parent in a mixed cytoplasm. This treatment facilitates the interspecific and intergeneric transfer of extranuclear genetic elements, such as mitochondria and chloroplasts, and is used to promote the transfer of genes which control chlorophyll content, herbicide resistance, and cytoplasmic male sterility. Cytoplasmic hybrids (cybrids) usually result from this procedure. Normally, cytoplasmic genomes are maternally inherited following sexual hybridization. Consequently, novel nucleo–cytoplasmic combinations can be produced sexually only by backcrossing, which is both time-consuming and unidirectional. Cybridization reduces this time period considerably.

2. Protoplast fusion

Protoplast fusion can be mediated by either chemical or electrical techniques. In both cases, the plasma membranes are temporarily destabilized, resulting in pore formation and cytoplasmic linkage between adjacent protoplasts. These linkages are thought to inhibit pore closure and allow randomly orientated lipid molecules in the pores to align and form membrane bridges between adjacent protoplasts.

In the case of chemical fusion, high concentrations of chemical fusogens such as polyethylene glycol (PEG), dextran, and polyvinyl alcohol (PVA) have been employed to induce pore formation, sometimes in combination with a high pH/Ca^{2+} buffer or a high pH/hypotonic buffer. Two steps are required for the electrofusion of protoplasts; first, an alternating current (AC) field is used to align the protoplasts and to initiate close membrane contact. Subsequently, a short direct current (DC) pulse is employed to induce membrane breakdown at the point of membrane contact.

Electrofusion is being used increasingly to mediate protoplast fusion since it is more reproducible and often less damaging to protoplasts than chemical procedures. However, an advantage of chemical procedures is that they avoid the use of sophisticated and expensive equipment which is required for the generation of AC fields and DC pulses.

2.1 Chemical fusion

Polyethylene glycol is by far the most commonly employed fusogen. In order to obtain a high frequency of heterokaryon formation (up to 10% of the treated protoplasts) and to ensure heterokaryon viability, it is essential to use PEG with a low carbonyl content, e.g. PEG of M_r 1500, supplied by Boehringer-Mannheim, or Aldrich Chemical Company, Ltd. Filter sterilize PEG solutions, since autoclaving also causes an increase in carbonyl content. Details for chemical fusion are given in *Protocol 1*.

2.2 Electrical fusion

The equipment necessary for electrically induced fusion consists of a function generator to produce the AC field, a DC source, a switching unit to apply the AC field or DC pulses, and a suitable fusion chamber. Commercial equipment is available (e.g. that marketed by Biotechnologies and Experimental

Research Inc., San Diego, or by B. Braun Biotech International GmbH,) as well as equipment built in laboratory workshops such as the versatile, low cost unit developed by the authors (1).

In preparation for electrical treatment, suspend the protoplasts, isolated as described in Chapter 2, in a medium of low conductivity, i.e. $10^{-4}-10^{-5}$ Ω/cm. The presence of ions in the medium should be avoided since these permit current flow. The latter reduces the strength of the AC field as well as causing heating and turbulence which reduce protoplast viability. Typical electro-fusion solutions consist of a non-ionic osmoticum (e.g. mannitol, sorbitol, sucrose, or glucose) in combination with small quantities of Ca^{2+} ions which help to stabilize the protoplast plasma membranes and to improve the efficiency of fusion. Reduce the osmotic pressure of the medium as low as possible without causing protoplast lysis, since this allows an increase in the protoplast diameter, leading to a greater voltage across the protoplast membrane.

In uniform AC electric fields, produced by parallel plate electrodes as used by most workers, the field strength on either side of the protoplasts should be equal. However, the presence of the protoplasts themselves disturbs the uniformity of the field, resulting in protoplast attraction and alignment along the field lines (formation of 'pearl chains'). Adjust the protoplast density to prevent long chain formation, but to permit the alignment of aggregates each consisting, ideally, of two protoplasts.

During electrofusion, protoplasts are aligned in an AC field of 0.5–2 MHz, 100–400 V/cm, for as short a time as is necessary to induce chain formation. The field strength is increased momentarily to induce close membrane-to-membrane contact before the DC pulse is applied to induce fusion. Ideally, the DC pulse should be applied while the AC field is still present. However, most equipment requires the AC field to be switched off before application of the DC pulse. The time between these two events should be minimal, since the interval without an AC field may be sufficient to cause separation of aggregated protoplasts. This separation can be induced by mutual repulsion of the protoplasts which carry a net negative surface charge. The critical DC voltage for membrane breakdown (V_m) in plants is of the order of 1 V per membrane (2). V_m is calculated by the equation

$$V_m = 1.5rE\cos\theta$$

in which r is the protoplast radius, θ is the angle between the normal to the membrane surface and the field line, and E is the electric field strength. Consequently, the voltage necessary to induce fusion is inversely proportional to protoplast size.

Other factors also affect the efficiency of fusion, particularly the condition of the plasma membranes following protoplast isolation, any remnants of undigested cell walls, and the presence of contaminating debris from burst protoplasts. The values for pulse intensity and duration necessary to induce fusion are also influenced by the degree of attraction and membrane contact resulting from the AC field. It is therefore necessary to determine the exact parameters at the time of each experiment (see *Protocol 2*). Application of multiple pulses of shorter duration is frequently preferable to a single long pulse.

Chemical treatments of protoplasts prior to electrofusion have been reported to both improve protoplast stability and increase fusion frequency (3, 4). The use of proteases (e.g. pronase and trypsin (5)), polyamines (spermine and/or spermidine (6)), and dimethylsulfoxide have been reported. However, care should be taken to ensure that such treatments do not impair protoplast viability. Proteases are the most effective enhancers of fusion frequency. Immediately before electrofusion, the protoplasts are incubated in a solution of pronase E (1 mg/ml) in CPW salts solution (7) with the same osmoticum as the electrofusion solution for 30 minutes, before being washed with electrofusion solution. Such treatments are beneficial if there is excessive protoplast bursting, if fusion frequencies are low, i.e. below 5%, or if there is a large size discrepancy between the protoplast fusion partners.

Protocol 1. Chemical fusion of protoplasts

The osmolarities described in this protocol are suitable for use with rice protoplasts, isolated as described in *Protocol 1*, Chapter 2. The mannitol concentrations may require adjustment for other protoplast combinations.

1. Make a 30% (w/v) solution of PEG 1500 (Boehringer-Mannheim UK, Lewes, UK) by adding aseptically 2.66 ml of buffer (17.8 g/litre Hepes, pH 8.0, autoclaved) to 4 ml of sterile 50% (w/v) PEG.

2. Prepare suspensions of both parental protoplasts as described in Chapter 2 and adjust their densities to 5×10^5/ml in 13% (w/v) mannitol solution in CPW salts (or whatever osmoticum is most suitable for the protoplasts); place in a refrigerator at 4°C for 10 min to lower the temperature.

3. Mix the protoplasts in equal quantities and dispense 10 ml into a 16 ml centrifuge tube.

4. Add 0.5 ml of 30% (w/v) PEG 1500, followed by 1.0 ml of hypotonic solution (9% (w/v) mannitol, 0.2% (w/v) bovine serum albumin, pH 5.8, filter sterilized, 4°C). Return the protoplasts to the refrigerator for 20 min.

5. Add 1.5 ml of hypotonic solution dropwise with gentle shaking, to ensure rapid mixing. Repeat this addition twice at 2 min intervals.

6. Add 4 ml of 11% (w/v) mannitol solution (pH 5.8); incubate at 4°C for 2 h in the refrigerator.

7. Centrifuge the protoplasts, resuspend in culture medium, and treat as described in Chapter 2.

Protocol 2. Electrofusion of protoplasts

This protocol describes the use of the electrofusion apparatus developed by the authors (1) which employs a parallel plate electrode system (8). Modifications should be made, as appropriate, for other equipment. Protoplasts can be observed during fusion using an inverted microscope inside a laminar air flow cabinet.

1. Disconnect the electrode from the apparatus and sterilize by immersion in 70% (v/v) ethanol solution for 10 min. Allow the electrode to dry inside the air stream of the flow cabinet.

2. Prepare suspensions of both parental protoplasts as described in Chapter 2 and transfer 1.5×10^6 protoplasts of each partner to separate centrifuge tubes. Add electrofusion solution, as described in Section 2.2 (typically 0.5 mM $CaCl_2$ with 11% (w/v) mannitol) to the top of each tube and centrifuge (15 min, 100 g).

3. Remove the supernatant from each tube and resuspend each protoplast pellet in 15 ml of electrofusion solution.

4. Dispense 0.75 ml of each protoplast preparation into each well of a 5 × 5 square well plastic Petri dish (Sterilin, Hounslow, UK), leaving two wells for protoplast self-fusion. Pipette 1.5 ml of the appropriate protoplast suspension into each of the self-fusion wells.

5. Switch on the oscilloscope, the AC function generator, and the DC supply. Select the required parameters:
   - AC frequency, 1 MHz
   - DC voltage, 1000 (V/cm)
   - number of pulses, 4
   - pulse width, 1.5 msec
   - inter-pulse separation, 1 sec

6. While observing the protoplasts, increase the AC amplitude and monitor pearl chain formation. Occasionally, spinning of the protoplasts is observed, in which case the AC frequency should be altered.

7. Increase the AC amplitude to maximum to induce close membrane contact between adjacent protoplasts and immediately activate the DC pulse sequence.

8. Reduce the AC field to the alignment amplitude, followed by a gradual reduction to zero.

9. Transfer the electrode to the next well and repeat the sequence, altering the DC parameters as necessary. If excessive bursting of the protoplasts is observed, the pulse width and DC voltage should be reduced. Conversely, these parameters should be increased if the fusion frequency

Protocol 2. *Continued*

was previously seen to be low (less than 5%). Frequently, large multiple fusions occur. This can be avoided by reducing the time of alignment in the AC field and by reducing the protoplast density.

10. Following fusion treatment, add 1.0 ml of culture medium to each well and leave the plate undisturbed for 1 h to allow the heterokaryons to stabilize. Subsequently, remove 1.0 ml of medium from each well and add 1.0 ml of fresh culture medium. After a further 15 min incubation at 22 °C, transfer the protoplasts to a centrifuge tube, mixing the contents of each well thoroughly to resuspend the protoplasts which have settled on to the base of the dish. After centrifugation and resuspension in medium, culture the protoplasts as described in Chapter 2.

3. Selection of heterokaryons and somatic hybrid tissues

The most difficult aspect of the somatic hybridization procedure is the selection of heterokaryons, heterokaryon-derived cells or tissues, or hybrid plants. Several approaches have been devised. Some workers have cultured all protoplasts subjected to the fusion treatment and screened the regenerated plants for their somatic hybrid or cybrid characteristics. The disadvantage of this procedure is that it is labour intensive and considerable growth room/glasshouse space may be required to house regenerated plants. Occasionally, heterokaryon-derived cells may exhibit heterosis and outgrow the cell colonies derived from parental protoplasts or homokaryons. Such faster growing colonies can be isolated mechanically and eventually transferred to plant regeneration medium. Other procedures which have been used successfully include manual selection of heterokaryons and high technology flow cytometry.

3.1 Manual selection

When the two parental protoplast populations can be easily identified, it is possible to pick out individual heterokaryons with micropipettes. This arrangement works well in the fusion of chlorophyll-containing leaf protoplasts with colourless protoplasts isolated from cell suspensions. A simple micromanipulation device has been reported (9). A major advance in this field has been the development of dual fluorescent labelling systems for heterokaryons. Protoplasts labelled green by treatment with fluorescein diacetate (1–20 mg/litre) are fused with protoplasts emitting a red fluorescence, either from chlorophyll autofluorescence (10), or from exogenously applied rhodamine isothiocyanate (10–20 mg/litre; (11)). Although micromanipulation of individual heterokaryons is both time-consuming and tedious, it is technically relatively straightforward and is a reliable method for somatic hybrid production (12).

3.2 Complementation

Procedures have been developed which inhibit growth of homokaryons and ideally both, but certainly one, of the parental protoplast populations. Mutants exhibiting hormone autotrophy, auxotrophy, resistance to anti-biotics, amino acid analogues, and fungal toxins have been used in various combinations (13,14), or with partners which have been inactivated by X- or γ-irradiation, or by treatment with iodoacetamide (15). Albino mutants have also been employed successfully and hybrid tissues identified by their chlorophyll production on exposure of cultures to light (16,17).

3.3 Flow cytometry

Flow cytometry can also be used to physically isolate heterokaryons (18). This technique uses the same method of identification of the heterokaryons, i.e. dual fluorescent labelling. Inside the flow cytometer, the protoplasts are passed in a flow of liquid between the light source and the fluorescence detectors. The stream is then dispersed into droplets and the computer electrostatically deflects the droplets containing heterokaryons into a sterile tube. The procedure is fully automatic and rapid; up to 2000 protoplasts per second can be analysed. However, the instrument is extremely costly and, for some protoplast systems, only about 10% of the heterokaryons are retained. This technique is normally used when it is possible to produce and fuse very large numbers of protoplasts ($> 5 \times 10^7$) of each parental population.

4. Characterization of somatic hybrid plants

Somatic hybrid plants have been characterized morphologically, cytologically, and biochemically. Routine biochemical characterization includes analyses of isozymes and fraction 1 protein, the resistance of plants to viral infection and antibiotics, or sensitivity to herbicides and fungal toxins. Genetic analysis can be performed, provided hybrid plants are fertile (14). More recently, the molecular techniques of RFLP and RAPD analyses have been employed (19–21), while flow cytometry provides a method for rapid analysis of the nuclear DNA content for ploidy determinations (22).

5. Concluding remarks

Protoplast fusion has an increasingly important rôle to play in plant biotech-nology. Reproducible protoplast-to-plant systems are now available for many plants of agronomic value, thus enabling protoplast fusion and transformation by direct gene uptake to be used as a means of genetic manipulation. Somatic hybrid and cybrid plants have been incorporated into breeding programmes and assessed under glasshouse/field conditions (23). Future work on protoplast fusion in major crops, such as rice, will be directed towards cybridization to increase yields through the transfer of characteristics such as male sterility and resistance to environmental stresses, including salt tolerance.

References

1. Jones, B., Lynch, P. T., Handley, G. J., Malaure, R. S., Blackhall, N. W., Hammatt, N., Power, J. B., Cocking, E. C., and Davey, M. R. (1994). *BioTechniques*, **6**, 312.
2. Zimmermann, U. and Scheruich, P. (1981). *Planta*, **151**, 26.
3. Nea, L. J., Bates, G. W., and Gilmer, P. J. (1987). *Biochim. Biophys. Acta*, **897**, 293.
4. Nea, L. J. and Bates, G. W. (1987). *Plant Cell Rep.*, **6**, 337.
5. Kameya, T. (1979). *Cytologia*, **44**, 449.
6. Chapel, M., Teissie, J., and Alibert, G. (1984). *FEBS Lett.*, **173**, 331.
7. Frearson, E. M., Power, J. B., and Cocking, E. C. (1973). *Dev. Biol.*, **33**, 130.
8. Watts, J. W. and King, J. M. (1984). *Biosci. Rep.*, **4**, 335.
9. Gilmour, D. M., Davey, M. R., and Cocking, E. C. (1987). *Plant Sci.*, **53**, 263.
10. Patnaik, G., Cocking, E. C., Hamil, J. D., and Pental, D. (1982). *Plant Sci. Lett.*, **24**, 105.
11. Barsby, T. L., Yarrow, S. A., and Shepard, J. F. (1984). *Stain Technol.*, **59**, 217.
12. Mendis, M. H., Power, J. B., and Davey, M. R. (1991). *J. Exp. Bot.*, **42**, 1565.
13. Potrykus, I., Jia, J., Lazar, G. B., and Saul, M. (1984). *Plant Cell Rep.*, **3**, 68.
14. Davey, M. R. and Kumar, A. (1983). *Int. Rev. Cytol. Suppl.*, **16**, 219.
15. Ozias-Akins, P., Ferl, R. J., and Vasil, I. K. (1986). *Mol. Gen. Genet.*, **203**, 365.
16. Cocking, E. C., George, D., Price-Jones, M. J., and Power, J. B. (1977). *Plant Sci. Lett.*, **10**, 7.
17. Gilmour, D. M., Davey, M. R., and Cocking, E. C. (1989). *Plant Cell Rep.*, **8**, 29.
18. Hammatt, N., Lister, A., Blackhall, N. W., Gartland, G., Ghose, T. K., Gilmour, D. M., Power, J. B., Davey, M. R., and Cocking, E. C. (1990). *Protoplasma*, **154**, 34.
19. Baird, E., Cooper-Bland, S., Waugh, R., Demaine, M., and Powell, W. (1992). *Mol. Gen. Genet.*, **233**, 469.
20. Polgár, Z., Preiszner, J., Dudits, D., and Fehér, A. (1993). *Plant Cell Rep.*, **12**, 399.
21. Xu, Y. S., Clark, M. S., and Pehu, E. (1993). *Plant Cell Rep.*, **12**, 107.
22. Hammatt, N., Blackhall, N. W., and Davey, M. R. (1991). *J. Exp. Bot.*, **42**, 659.
23. Grosser, J. W., Gmitter, F. G., Louzada, E. S., and Chandler, J. L. (1992). *Hortscience*, **27**, 1125.

3B. Transient gene expression and stable transformation

R. A. DIXON

1. Introduction

The removal of the cell wall facilitates the uptake of foreign DNA into plant cells. The most commonly used technique for DNA delivery to protoplasts is that of electroporation, in which application of high voltage DC pulses is believed to open up pores in cell membranes through which nucleic acids can migrate by diffusion or electro-osmosis (1, 2). Originally, apparatus used for electroporation was home-made, but a number of electroporators suitable for use with plant protoplasts are now commercially available. These include the Promega Biotech X-Cell 450 and ZA 2010 models (Prototype Design Services, Madison, WI), the Baekon 2000 Advanced Gene Transfer System (Baekon, Saratoga, CA), the Gene-Pulser^R (Bio-Rad, Hercules, CA, USA), and the BTX-Transfector 300 (BTX, San Diego, CA, USA). The voltage and pulse length required for optimal DNA transfer via electroporation vary according to the source of protoplasts, the electroporation medium, and the type of pulse (square wave or exponential wave). Protoplast viability is invariably decreased as a result of electroporation, and it is necessary to determine experimentally the conditions which give the best compromise between DNA uptake and viability. The reader should consult the references listed in *Tables 1* and *2* in order to decide on initial electroporation conditions for various plant species.

Simple exposure of plant protoplasts to polyethylene glycol (up to a maximum of 28%, v/v) in the presence of a high salt medium ($MgCl_2$) can result in DNA uptake (3–5). PEG of molecular weight 4000 to 8000 is usually used. Some procedures utilize electroporation in the presence of PEG (6–8).

2. Transient gene expression

Electroporation of DNA constructs into plant protoplasts is a powerful tool for the functional analysis of plant gene promoters. In essence, the promoter fragment being analysed is fused to a suitable reporter gene (chloramphenicol acetyltransferase (CAT) and β-glucuronidase (GUS) being the most popular) and an efficient terminator sequence, and reporter gene activity is assayed enzymatically after introduction of the construct into protoplasts followed by a suitable incubation period. It is also possible to measure directly the transcripts produced from the introduced gene constructs (9). *Table 1* summarizes a cross-section of the literature on transient gene expression in plant

Table 1. Selected examples of transient gene expression in plant protoplasts

Source of protoplasts	Promoter[a]	Reporter gene[b]	Method[c]	Reference
Alfalfa cell suspension	Bean chs	cat	E + PEG	6
Arabidopsis thaliana cell suspension	CaMV 35S, Arabidopsis EF-α	gus	PEG	10
Bean (P. vulgaris) cell suspension	CaMV 35S + TMVΩleader	luc, gus	E + PEG	7
Bean (P. vulgaris) cotyledons	Bean phaseolin	gus, cat	E	11
Black spruce	CaMV 35S, nos	cat	E	12
Carrot cell suspension	nos	cat	E	1
Carrot cell suspension	None (RNA used)	cat (transcript)	E	13
Jack pine	CaMV 35S, nos	cat	E	12
Maize cell suspension	None (RNA used)	cat (transcript)	E	13
Maize cell suspension	CaMV 35S, maize adhl	luc	E	14
Nicotiana plumbaginifolia cell suspension	Rauvolfia serpentina strl	gus	E	15
Nicotiana plumbaginifoia leaves	Soybean GMhsp17.5 E	gus	E	16
Parsley cell suspension	Antirrhinum majus chs	nptll	PEG	4
Rice cell suspension	CaMV 35S	cat	E	17
Sorghum callus culture	CaMV 35S	cat	E	17
Soybean cell suspension	Bean chs	cat	E	18
Soybean cotyledons	CaMV 35S	cat	E + PEG	8
Tobacco leaf mesophyll	Tobacco par	gus	E	19
Wheat cell suspension	CaMV 35S	cat	E	17

[a] adh, alcohol dehydrogenase; CaMV, cauliflower mosaic virus; chs, chalcone synthase; EF, elongation factor; Gmhsp, Glycine max heat shock protein; nos, nopaline synthase; par, protoplast auxin regulated; str, strictosidine synthase; TMV, tobacco mosaic virus.
[b] cat, chloramphenicol acetyltransferase; gus, β-glucuronidase; luc, luciferase; npt, neomycin phosphotransferase.
[c] E = electroporation, PEG = treatment with polyethylene glycol.

protoplasts. Note that a selectable marker is not employed, as stable transformation is not being sought. *Protocols 1* and *2* describe methods we have utilized for analysis of constructs driven by a bean chalcone synthase promoter and various deletions and mutations thereof (18, 22). Be sure that, if cell suspensions are to be used as protoplast source, the suspensions are growing rapidly and the cells are as small and homogeneous as possible. Slow growing, unheathly suspensions generally yield protoplasts which do not survive electroporation well.

Protocol 1. Determination of transient gene expression by electroporation in soybean protoplasts

1. Maintain soybean suspension cultures in MS medium (see Chapter 1) supplemented with 0.5 mg/litre 2,4-D and 0.5 mg/litre 6-benzylaminopurine. Maintain a rapidly growing culture by subculturing every seven days after collecting cells on a 250 μm nylon mesh sieve.

2. Prepare and filter sterilize the protoplasting enzyme solution which contains 1% (w/v) cellulase (Cellulysin, Calbiochem-Behring), 0.5% hemicellulase (Rhozyme, Genencor), 0.2 M mannitol, 50 mM $CaCl_2$, 10 mM sodium acetate, pH 5.8.

3. Collect cells (7 g fresh weight, four days after subculture), transfer to 100 ml of protoplasting enzyme solution in a 250 ml conical flask, and incubate with shaking (90 r.p.m.) at 27 °C in the dark for 4 h.

4. Separate protoplasts from large debris by filtration through a 60 μm mesh nylon screen and centrifugation at 70 *g* for 5 min at room temperature. Wash protoplasts once in 0.2 M mannitol, 50 mM $CaCl_2$, 10 mM sodium acetate, pH 5.8.

5. Count the protoplasts in a haemocytometer and determine viability using Evan's blue stain (see Chapter 1).

6. Resuspend protoplasts at a density of 5×10^6/ml, then wash twice in electroporation medium[a] by gentle centrifugation (70 *g*, 5 min, room temperature).

7. Electroporate 600 μl batches of protoplasts on ice in 1 ml plastic spectrophotometer cuvettes containing the DNA construct (30 μg) plus 50 μg carrier calf thymus DNA. Electroporation conditions will vary depending on the origin of the protoplast preparation; try 250 V for 10 msec as initial conditions, and then vary voltage and pulse length. Consult ref. 1 for a consideration of the effects of voltage and pulse length on protoplast viability and subsequent gene expression.

8. Incubate protoplasts on ice for 10 min then at room temperature for 10 min.

Protocol 1. *Continued*

9. Slowly dilute protoplasts with 6 ml MS medium containing 2% (w/v) sucrose, 0.3 M mannitol and 0.1 μg/ml 2,4-D and incubate, without stirring, in Petri dishes in the dark at 27 °C.

10. For inducible genes, treat protoplasts with inducing agent at various times up to 20 h after preparation and subsequently harvest for assay of reporter gene activity.

[a] Electroporation medium is 10 mM Hepes, pH 7.2, 150 mM NaCl, 5 mM $CaCl_2$, 0.2 M mannitol.

Protocol 2. Transient gene expression in alfalfa protoplasts: electroporation plus PEG method

1. Initiate suspension cultures of alfalfa (*Medicago sativa* L.) from callus cultures in modified SH medium (see Chapter 1) supplemented by 1.8 mg/litre *p*-chlorophenoxyacetic acid, 0.5 mg/litre 2,4-D, 0.5 mg/litre kinetin, 100 mg/litre L-serine, 800 mg/litre L-glutamine, and 1 mg/litre adenine. Grow in the dark at 25 °C with constant shaking and subculture (10 ml culture into 40 ml fresh medium) every seven days.

2. Prepare and filter-sterilize the protoplasting enzyme solution which contains 1% (w/v) driselase (Sigma), 1% (w/v) cellulase (Onozuka RS), 0.5% (w/v) macerozyme (Onozuka R10), 0.5% (w/v) hemicellulase (Sigma), and 0.4 M mannitol in suspension culture medium, pH 5.8.

3. Incubate cultured cells (1 g, five to six days after subculture) with 10 ml protoplasting enzyme solution in 100 × 15 mm Petri dishes with shaking (40 r.p.m.) at 25 °C in the dark for 12–14 h.

4. Separate protoplasts from undigested cells and debris by successive passages through 70, 40 and 30 μm nylon mesh filters. Collect protoplasts by centrifugation at 100 *g* for 10 min, and wash three times with W5 medium, pH 5.8.[a]

5. Count the protoplasts in a haemocytometer and determine viability using Evan's blue stain (see Chapter 1). The yield should be approximately 1×10^6/g fresh weight, with viability in excess of 90%, and no cell wall residues should be present as observed by Calcofluor White staining (20). If these criteria are not met the protoplasts should not be used.

6. Resuspend protoplasts at a density of 2×10^7/ml in MsMg solution.[b]

7. Heatshock the protoplasts for 5 min at 45 °C and then bring to room temperature by incubating on ice for 2 min.

8. Transfer 500 μl aliquots of heatshocked protoplasts to 1 ml plastic spectrophotometer cuvettes containing the DNA to be electroporated (10–60 μg plasmid containing the gene promoter construct plus 50 μg carrier calf thymus DNA). Incubate at room temperature for 10 min.

9. Add 200 μl of 40% PEG solution [c] and incubate on ice for 10 min.

10. Electroporate using a commerically available electroporation apparatus such as a BTX Transfector 300 unit (BTX, San Diego).

11. Incubate protoplasts on ice for 10 min, then at room temperature for 10 min, before diluting with 5 vol. 0.2 M $CaCl_2$, 0.4 M mannitol. Centrifuge for 5 min at 50 g, wash with protoplast culture medium (21), and culture in 2 ml of protoplast culture medium in the dark at 25 °C.

12. For inducible genes treat protoplasts with inducing agent (e.g. elicitors, hormones, UV light) at various times up to 20 h after preparation and subsequently harvest for assay of reporter gene activity.

[a] W5 medium is 154 mM NaCl, 125 mM $CaCl_2$, 5 mM KCl, 5 mM glucose, 0.4 M mannitol, pH 5.8–6.0.
[b] MsMg solution is Murashige and Skoog salts (see Chapter 1), 0.4 M mannitol, 30 mM $MgCl_2$, 0.1% (w/v) MES, pH 5.8.
[c] 40% PEG solution is 40% (v/v) polyethylene glycol (M_r 8000), 0.1% (w/v) MES, 0.4 M mannitol, 30 mM $MgCl_2$, pH 7.0.

If you are interested in analysing an inducible promoter, it is obviously essential that the protoplasts retain responsiveness to the inducing agent. Protoplast electroporation has been successfully used to analyse UV-, auxin-, heat-, fungal elicitor-, and glutathione-responsive gene promoters. The act of preparing protoplasts can result in elicitation of defence response genes, thereby making it difficult to study elicitor-responsiveness. In soybean cell suspension protoplasts, the elicitation by protoplasting is transient, and the ratio of inducible to basal expression of defence response genes increases to an acceptable level within approximately 12 hours after protoplasting (18). Parsley protoplasts retain their responsiveness to UV light and fungal elicitors with no basal elicitation as a result of protoplasting (23).

3. Stable transformation

Introduction of double stranded DNA into protoplasts by electroporation and/or PEG treatment can result in stable chromosomal integration of a small fraction of the DNA, as initially assessed by selection of microcalli derived from the protoplasts on a medium containing an antibiotic or antimetabolite for which a gene conferring resistance is incorporated in the introduced DNA. A typical transformation frequency for tobacco protoplasts is approximately 1400 transformants per 3×10^5 treated protoplasts using the PEG/ $MgCl_2$ method; frequencies for other species may be considerably lower (3). Single stranded DNA has been shown to give a three- to ten-fold higher frequency of stable transformation than double stranded DNA for tobacco (24).

The key to successful plant transformation by direct DNA transfer into protoplasts is the ability to regenerate plants from the protoplast-derived microcalli. *Table 2* summarizes the literature on stable transformation of

Table 2. Selected examples of stable transformation using plant protoplasts

Source of protoplasts	Promoter[a]	Selectable marker[b]	Method[c]	Outcome[d]	Reference
Maize cell suspension	CaMV 35S	nptII	E or PEG	ST	25
Maize cell suspension	CaMV 35S	nptII	E	ST	26
Maize cell suspension	CaMV 35S	nptII	E	ST + R	27
Orchardgrass cell suspension	CaMV 35S	hph	E or PEG	ST + R	28
Panicum maximum cell suspension	CaMV 35S	dhfr	E	ST	29
Rice cell suspension	CaMV 35S	hph	E	ST + R	30
Rice cell suspension	CaMV 35S	nptII	E	ST + R	31
Soybean cotyledons	CaMV 35S	nptII	E	ST + R	32
Tall fescue cell suspension	CaMV 35S	hph, bar	PEG	ST + R	5
Tobacco mesophyll	nos	nptII	L	ST + R	33
Tobacco shoot cultures	CaMV 19S, nos	nptII	PEG	ST	3
Triticum monococcum cell suspension	CaMV 35S	hph	E	ST	29

[a] CaMV, cauliflower mosaic virus; nos, nopaline synthase.
[b] npt, neomycin phosphotransferase; hph, hygromycin phosphotransferase; bar, phosphinothricin acetyltransferase; dhfr, dihydrofolate reductase.
[c] E = electroporation, PEG = polyethylene glycol treatment, L = liposome-mediated DNA delivery.
[d] ST = stable transformation of cell culture, ST + R = stable transformation and plant regeneration.

plant protoplasts; note that regeneration was not always attempted or achieved. This technology has been widely applied to gramineous species, which are recalcitrant to *Agrobacterium*-mediated stable transformation.

References

1. Fromm, M., Taylor, L. P., and Walbot, V. (1985). *Proc. Natl. Acad. Sci. USA*, **82**, 5824.
2. Van Wert, S. L. and Saunders, J. A. (1992). *Plant Physiol.*, **99**, 365.
3. Negrutiu, I., Shillito, R., Potrykus, I., Biasini, G., and Sala, F. (1987). *Plant Mol. Biol.*, **8**, 363.
4. Lipphardt, S., Brettschneider, R., Kruezaler, F., Schell, J., and Dangl, J. L. (1988) *EMBO J.*, **7**, 4027.
5. Wang, Z., Takamizo, T., Iglesias, A., Osusky, M., Nagel, J., Potrykus, I., and Spangenberg, G. (1992). *Bio/Technology*, **10**, 691.
6. Choudhary, A. D., Kessmann, H., Lamb, C. J., and Dixon, R. A. (1990). *Plant Cell Rep.*, **9**, 42.
7. Leon, P., Planckaert, F., and Walbot, V. (1991). *Plant Physiol.*, **95**, 968.
8. Lin, W., Odell, J. T., and Schreiner, R. M. (1987). *Plant Physiol.*, **84**, 856.
9. Murray, E. E., Buchholz, W. G., and Bowen, B. (1990). *Plant Cell Rep.*, **9**, 129.
10. Axelos, M., Curie, C., Mazzolini, L., Bardet, C. and Lescure, B. (1992). *Plant Physiol. Biochem.*, **30**, 123.
11. Bustos, M. M., Battraw, M. J., Kalkan, F. A., and Hall, T. C. (1991). *Plant Mol. Biol. Rep.*, **9**, 322.
12. Tautorus, T. E., Bekkaoui, F., Pilon, M., Datla, R. S. S., Crosby, W. L., Fowke, L. C., and Dunstan, D. I. (1989). *Theor. Appl. Genet.*, **78**, 531.
13. Callis, J., Fromm, M., and Walbot, V. (1987). *Nucleic Acids Res.*, **15**, 5823.
14. Planckaert, F. and Walbot, V. (1989). *Plant Cell Rep.*, **8**, 144.
15. Bracher, D. and Kutchan, T. M. (1992). *Arch. Biochem. Biophys.*, **294**, 717.
16. Ainley, W. M. and Key, J. L. (1990). *Plant Mol. Biol.*, **14**, 949.
17. Ou-Lee, T. M., Turgeon, R., and Wu, R. (1986). *Proc. Natl. Acad. Sci. USA*, **83**, 6815.
18. Dron, M., Clouse, S. D., Dixon, R. A., Lawton, M. A., and Lamb, C. J. (1988). *Proc. Natl. Acad. Sci. USA*, **85**, 6738.
19. Takahashi, Y., Niwa, Y., Machida, Y., and Nagata, T. (1990). *Proc. Natl. Acad. Sci. USA*, **87**, 8013.
20. Nagata, T. and Takebe, I. (1970). *Planta*, **92**, 12.
21. Menczel, L., Nagy, F., Kiss, Z., and Maliga, P. (1981). *Theor. Appl. Genet.*, **59**, 191.
22. Harrison, M. J., Choudhary, A. D., Dubery, I., Lamb, C. J., and Dixon, R. A. (1991). *Plant Mol. Biol.*, **16**, 877.
23. Dangl, J. L., Hauffe, K. D., Lipphardt, S., Hahlbrock, K., and Scheel, D. (1987). *EMBO J.*, **6**, 2551.
24. Rodenburg, K. W., de Groot, M. J. A., Schilperoort, R. A., and Hooykaas, P. J. J. (1989). *Plant Mol. Biol.*, **13**, 711.
25. Lyznik, L. A., Kamo, K. K., Grimes, H. D., Ryan, R., Chang, K. L., and Hodges, T. K. (1989). *Plant Cell Rep.*, **8**, 292.

26. Fromm, M. E., Taylor, L. P., and Walbot, V. (1986). *Nature*, **319**, 791.
27. Rhodes, C. A., Pierce, D. A., Mettler, I. J., Mascarenhas, D., and Detmer, J. J. (1988). *Science*, **240**, 204.
28. Horn, M. E., Shillito, R. D., Conger, B. V., and Harms, C. T. (1988). *Plant Cell Rep.*, **7**, 469.
29. Hauptmann, R. M., Vasil, V., Ozias-Akins, P., Tabaeizadeh, Z., Rogers, S. G., Fraley, R. T., Horsch, R. B., and Vasil, I. K. (1988). *Plant Physiol.*, **86**, 602.
30. Shimamoto, K., Terada, R., Izawa, T., and Fujimoto, H. (1989). *Nature*, **338**, 274.
31. Toriyama, K., Arimoto, Y., Uchimiya, H., and Hinata, K. (1988). *Bio/Technology*, **6**, 1072.
32. Dhir, S. K., Dhir, S., Sauka, M. A., Belanger, F., Kriz, A. L., Farrand, S. K., and Widholm, J. M. (1992). *Plant Physiol.*, **99**, 81.
33. Deshayes, A., Herrera-Estrella, L., and Caboche, M. (1985). *EMBO J.*, **4**, 2731.

3C. Studies with viruses

JAMES C. REGISTER III

1. Introduction

Since the first report of *in vitro* inoculation of protoplasts with a virus almost a quarter of a century ago (1), this system has been used extensively for examination of both virus replication cycles and the functions encoded by virus genomes. Over the past several years a number of new techniques and applications have been introduced using this system. This section provides an update of techniques not covered by Wood (2) in the first edition of this book, and focuses on the use of protoplasts for studying virus resistance.

2. Appropriate uses/limitations of protoplasts

There are two principle advantages of using protoplasts rather than whole plants in virus infection studies: in the absence of virus spread one is assured that all replication is in primary infected cells, and the achievable synchrony of infection permits dissection of replication (particularly early events) with a precision impossible with whole plants.

These advantages must always be weighed against the potential pitfalls of using protoplasts. First, there will always be a minority who believe that since protoplasts reflect an artificial situation, experiments using them are inherently artefact-ridden. Although countless studies using protoplasts have proven this view false, caution must be exercized in order to safeguard against artefacts and to interpret results properly. One valuable safeguard is to support protoplast experiments with experiments utilizing intact plants; however this is clearly not always possible. Considering the following questions before initiating a project using protoplasts can also help avoid many problems:

(a) What is the proper tissue source for the protoplasts? This question can be important either because the virus shows tissue-specificity in plants or because early events of infection need to be specifically addressed.

(b) Is insect transmission of the virus a compounding factor?

(c) If inoculation with virions is required, is there any evidence that infection is difficult to achieve? For instance, infection by potyviruses (virions, but not RNA) has been notoriously difficult to achieve (3,4).

3. Protoplast preparation

For the vast majority of virus replication studies that used protoplasts in the past, the tissue source of the protoplasts was a secondary concern—any tissue from a host species that provided sufficient protoplast yields generally sufficed. As protoplasts begin to be used in more sophisticated ways, however, tissue source may become a critical part of experimental design due to such factors as outlined above.

Numerous papers exist which describe isolation of protoplasts from different species and tissues, and an overview of the methods used has been provided in Chapter 2. For studies of virus resistance, one decision may be whether leaf mesophyll or suspension culture protoplasts (which are typically both relatively easy to obtain), or epidermal protoplasts (typically more difficult to obtain in large numbers), are most appropriate. It is important to note that the conditions used for protoplast isolation from just these three sources can be quite different, even from a single species, as shown for tobacco in *Table 1*.

4. Inoculation

One of the more significant developments in this field since publication of the first edition of this book has been the use of electroporation for inoculation of protoplasts with virus or viral nucleic acid. The reproducibility of electroporation and polyethylene glycol (PEG)-mediated inoculation, together with a variety of difficulties with other techniques such as poly-L-ornithine- or liposome-mediated inoculation, has made the first two approaches the methods of choice for protoplast inoculation. PEG-mediated inoculation was dealt with in detail by Wood (2).

4.1 Electroporation

Despite the widespread use of electroporation, it is impossible to provide a single optimal protocol, due to the variety of electroporators, buffers, and

Table 1. Comparison of conditions used for isolation of protoplasts from three tobacco tissue sources

	Mesophyll	Epidermal	Suspension
Isolation medium	0.5 M mannitol, 10 mM CaCl$_2$, 5 mM MES (pH 5.7) (0.1% potassium dextran sulfate, 0.5% BSA)	0.3 M mannitol, 10 mM CaCl$_2$, 10 mM MES (pH 5.5), 1% BSA	0.4 M mannitol, 10 mM CaCl$_2$, 5 mM MES (pH 5.7)
Enzymes	Cellulase R10—1.5% Macerozyme R10—0.15%	Cellulase R10—1% Pectolyase—0.005%	Cellulase R10—3% Macerozyme R10—0.2% Driselase—0.2%
Time of digestion	4–6 h at 25 °C	15–30 min at 20–25 °C	Approx. 14 h at 25 °C
Flotation step	Yes, 23% sucrose	Yes, 15% Percoll	Optional, if needed— 23% sucrose
Incubation medium	KH$_2$PO$_4$—0.2 mM, KNO$_3$—1 mM, MgSO$_4$—1 mM, CaCl$_2$—10 mM, KI— 1 μM, CuSO$_4$—0.01 μM, Mannitol—0.5 M, 2,4-D— 1 mg/litre, Carbenicillin— 200 μg/ml, (Mycostatin— 5 μg/ml), pH 5.6	KH$_4$PO$_4$—0.4 mM, KNO$_3$—2 mM, MgSO$_4$— 0.2 mM, CaCl$_2$—0.2 M, KI—2 μM, CuSO$_4$— 0.02 μM, D-Glucose—25 mM, Cephaloridine—100 μg/ml, Carbenicillin—100 μg/ml, Mycostatin—4 μg/ml, pH 5.5	See mesophyll
Reference	5, 6	7, 8	

conditions that different researchers have used. This topic has been recently reviewed (9). *Table 2* provides a comparison of parameters used successfully by a variety of researchers. *Protocol 1* has been successfully used with tobacco leaf mesophyll protoplasts and tobacco mosaic virus (TMV) or TMV-RNA (5).

Protocol 1. Electroporation of tobacco mesophyll protoplasts with TMV or TMV-RNA

1. Wash protoplasts two to three times in ice-cold electroporation buffer (70 mM KCl, 5 mM MES, pH 5.7, same mannitol concentration as in isolation buffer) (13).
2. Resuspend protoplasts at 2.5 × 10^6 cells/ml, hold on ice.
3. Add desired amount of inoculum to cells in a minimal volume.
4. Transfer 1 ml of evenly resuspended protoplasts to electroporation cuvette (this volume will be dependent upon the cuvettes used).
5. Electroporate with a single exponential decay discharge of 500 V/cm, 320 μF.
6. Gently transfer cells to ice, hold for ≥ 10 min prior to diluting in incubation medium (see *Table 1*).

Table 2. Comparison of electroporation parameters used to successfully inoculate protoplasts with virus or viral nucleic acid

Protoplast source	Electroporator[b]	Buffer[a]	Cells/ml	Discharge	Inoculum[b]	Reference
Tobacco mesophyll	HM/ED	70 mM KCl 5 mM MES (5.7)	2.5×10^6	1×500 V/cm 320 µF	TMV, TMV-RNA	5
Tobacco mesophyll	Bio-Rad/ED	150 mM KCl 1 mM KPO$_4$ (7.0)	1×10^6	1×250 V/cm 1000 µF	TVMV-RNA	10
Tobacco mesophyll	HM/ED	150 mM KCl 1 mM KPO$_4$ (7.0)	3×10^6	1×250 V/cm 1000 µF	TVMV-RNA PVY-RNA	3
Tobacco suspension, N. clevelandii mesophyll	HM/ED	150 mM NaCl 4 mM CaCl$_2$ 10 mM MES (5.7)	$1–3 \times 10^6$	1×600 V/cm 450 µF	RCNMV RCNMV-RNA	11
Tobacco or N. plumbaginofolia mesophyll	HM/ED	No electrolytes	$0.5–1 \times 10^6$	1×400 V/cm 100 µF or 1×1500 V/cm, 50 nF	CCMV-RNA, BMV	12
Tobacco, V. rosea, rice suspension	HM/ED	70 mM KCl 5 mM MES (5.8)	3×10^6	1×750 V/cm 100 µF	CMV-RNA, TMV-RNA	13, 14
Tobacco mesophyll	GCA, SW	No electrolytes	$1–1.5 \times 10^6$	$9 \times 5–10$ kV/cm 90 µsec	TMV	15
Tobacco mesophyll	Sankei Co, SW	No electrolytes	$1–3 \times 10^6$	1×670 V/cm 100 µsec	TMV-RNA, CMV-RNA	16
Tobacco mesophyll	HM/ED	No electrolytes	2×10^5	$5 \times 600–800$ V/cm	TMV-RNA	17
Tobacco mesophyll	BTX/ED or SW	No electrolytes	5×10^5	1×530 V/cm 22 msec (ED) 1×2000 V/cm 80 µsec (SW)	CMV-RNA	18

[a] Osmotica are not given, values in parentheses are pH.
[b] HM, home-made; ED, exponential decay; SW, square wave; GCA, GCA Corp., Chicago, IL; BTX, BTX, Inc., San Diego, CA; TVMV, tobacco vein mottling virus; PVY, potato virus Y; RCNMV, red clover necrotic mottle virus; CCMV, cowpea chlorotic mottle virus; BMV, brome mosaic virus; CMV, cucumber mosaic virus.

When incubating protoplasts in liquid medium within plastic Petri dishes, cells will often stick to the plates. While this can be at least partially avoided by coating the plates with a thin film of 1% agar (12), the per cent viability is typically higher when these cells are allowed to stick, indicating that these cells are dead or dying.

5. Assessing virus accumulation

5.1 Removal of cross-reacting antibodies

Many of the techniques used for assessing virus accumulation require the use of antibodies. Although antibodies against any viral product can be used to monitor infection, only antibodies against a structural protein (e.g. the coat protein; CP) can be used to monitor virus accumulation. One potential problem that must be dealt with prior to carrying out large scale experiments is the presence of cross-reacting antibodies in either primary or secondary antibody preparations. For techniques such as western immunoassays, cross-reactivity is often a minor nuisance, but for others, such as fluorescent antibody staining, dot-blots, or ELISA, cross-reactivity can result in unacceptably high backgrounds. *Protocol 2* (established for tobacco leaves) was originally published by Hibi and Saito (19); an alternative approach is the use of glutaraldehyde-activated resins (such as that sold by Boehringer-Mannheim, Product Number 665 525) to which either plant protein preparations, or purified virus, can be bound.

Protocol 2. Removal of cross-reacting antibodies from anti-(viral) sera

A. *Preparation of leaf powder*

1. Freeze healthy tissue (250 g) at $-70\,^{\circ}$C and store if desired.

2. Grind tissue in 250 ml of 0.1 M NaPO$_4$ buffer (pH 7.2), 0.1% thioglycolic acid, then centrifuge at 8000 g for 15 min at 4$\,^{\circ}$C.

3. Filter supernatant through cheesecloth and make solution 4% (w/v) with respect to PEG (M_r 6000) and 0.2 M NaCl. Hold on ice for 30 min.

4. Collect precipitate by centrifugation as in step **2** above. Resuspend in 100 ml of 0.1 M NaPO$_4$ buffer (pH 7.2), 10 mM EDTA.

5. Centrifuge at 10 000 g for 10 min, save supernatant.

6. Centrifuge supernatant at 10 000 g for 1 h. Freeze-dry pellet and pulverize prior to use.

B. *Adsorption of antibodies*

1. Dilute antibody to working strength in PBS (10 mM NaPO$_4$ pH 7.2, 0.15 M NaCl), add tobacco powder at 10 mg/ml of antibody solution.

---

2. Stir for 2 h at 37 °C.

3. Centrifuge at 15 000 g for 10 min, save supernatant, then repeat steps **5** and **6** two times using the powder at 5 mg/ml.

4. Store antibody solution as usual until use.

5.2 Protein dot-blot assay

Although ELISA provides the most easily quantifiable assay for viral protein accumulation, handling large numbers of samples can become burdensome in the absence of automated equipment. One alternative is a quantitative protein dot-blot assay (5). *Protocol 3* is modified from that of Hibi and Saito (19).

Protocol 3. Quantitative dot-blot immunoassay

1. Collect protoplasts as a pellet and store at −70 °C.

2. Resuspend at a concentration equivalent to 2500 cells/μl in 0.1 M NaPO$_4$ (pH 7.2).[a]

3. To ensure that antigen concentrations are within the range of standards used for the assay, prepare a series of twofold dilutions of the protoplasts starting at 2500 cells/μl (or 500 ng of total protein/μl). These samples will be used in the assay. If quantitation is desired it is also important to include a series of antigen (typically CP) standards diluted in healthy plant extract.

4. If a commercial dot-blot apparatus is to be used, follow the manufacturer's instructions—if one is not available, continue as outlined below.

5. Cut a piece of nitrocellulose membrane to the desired size and mark off a 1 cm grid pattern.[b]

6. Soak the membrane in PBS for 15 min, then dry between sheets of filter paper for 5 min.

7. Spot 2 μl samples at the centre of grid squares.[c]

8. Air dry the membrane for 5 min, then bake at 65 °C for 15 min.

9. Treat as for standard western blot immunoassay.

---

[a] The assay can be run on either a per cell or total protein basis. The former may be advantageous for correlation with other assays (see Section 6) while total protein can be determined using the method of Bradford (20).
[b] Other membranes (nylon, PVDF) can also be used and may give superior results due to binding characteristics and tensile strength.
[c] A mechanical micropipettor may be used for this; however, even greater accuracy can be obtained using a Hamilton syringe with a long piece of flexible plastic tubing (such as Tygon) attached to the needle. Contaminated tubing can be cut off following application of each sample.

Secondary antibody conjugated to ^{125}I is very useful in the dot-blot assay as results can be obtained either on autoradiographs, which can be scanned, or by counting individual samples in a gamma-counter. A chemiluminescent secondary antibody system can also be used for detection on film, eliminating the need for radioisotopes.

5.3 Nucleic acid replication

Several options are available for detection of nucleic acid replication, and the choices will vary depending on the genome of the virus (i.e. (+) or (−) strand RNA, double strand (ds) RNA, single strand or dsDNA). Wood (2) has previously discussed the use of probes made by reverse transcription of RNA genomes. These probes provide specific detection of (+)strand accumulation for (+)strand RNA viruses (which comprise > 75% of all plant viruses). Recently, the polymerase chain reaction (PCR) has opened up a variety of alternative approaches for probe synthesis. Although there is not space here to deal with PCR in detail, reviews and protocols have been published (e.g. 21,22). For an RNA virus such as TMV, reverse transcription can be coupled with PCR to yield dsDNA. Subsequent rounds of standard or asymmetric PCR can yield DNA suitable for either general or strand-specific probes. Probes can be produced directly during the PCR reaction or in subsequent reactions such as random oligonucleotide-primed labelling (21).

6. Applications to resistance

One area of virology in which the use of protoplasts is likely to expand over the next several years is the study of virus resistance. Although many resistances are due to blockages of virus spread, others clearly act at the single cell level. Examples include natural, induced, and genetically engineered forms of resistance (see ref. 5 and references therein).

To provide an example of how to dissect such a resistance systematically, the approach used for studying coat protein-mediated protection (CPMP) against TMV in transgenic tobacco plants is presented. Expression of a chimeric TMV CP gene in these plants (CP (+)) confers resistance to TMV infection. This system is a useful model for two reasons. First, because CPMP against TMV acts by blocking a very early event in virus infection, the experimental strategy addresses several steps in replication. Secondly, these studies emphasize the importance of understanding the replication cycle of the virus being studied. A very simplified diagram of the TMV replication cycle is presented in *Figure 1*.

As this section is not a review of CPMP, most CPMP-specific references have been omitted—all this information can be found in recent reviews (23, 24).

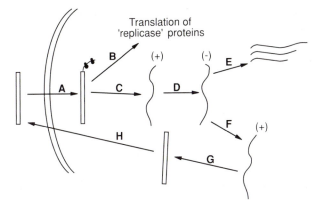

Figure 1. Replication cycle of TMV. TMV enters a host cell and co-translational disassembly is initiated (A). The ultimate products are the putative replicase proteins (B) and genomic RNA (C), which is replicated to yield (−)strand RNA (D). This RNA is replicated to yield either subgenomic RNAs (E), which are templates for translation of other viral proteins (including CP), or genomic RNA (F) which is encapsidated (G) to produce infectious virions (H).

6.1 Resistance at the single cell level

For studies of the mechanism of resistance at the single cell level, one should initially evaluate protection by assessing accumulation of a late viral product following inoculation of protoplasts from susceptible and resistant plants. As for most viruses, CP accumulation is the logical choice for TMV infection (see *Figure 1*).

For evaluation of CPMP, inoculate protoplasts from transgenic CP(+) and CP(−) plants with TMV. Two complementary methods have been used for quantifying infection: fluorescent antibody staining and a quantitative dot immunobinding assay (see *Protocol 3*). Although the results obtained with either assay clearly demonstrate protection (see *Figure 2*), the combination of assays provides useful additional information. By extrapolating from the data shown in *Figure 2*, it can be seen that the few infected CP(+) protoplasts contain approximately the same amount of virus as infected control cells, indicating that protection appears to be an 'all-or-nothing' process. In these experiments it is important to inoculate with increasing concentrations of virus as it has been shown that protection in plants can be overcome by high inoculum concentrations. The results shown in *Figure 2* indicate that a progressive breakdown of protection does not occur at the single cell level.

6.2 Blockage of nucleic acid replication

Nucleic acid replication precedes CP accumulation in TMV infected cells (see *Figure 1*). Nucleic acid dot-blots (2) provide the most straightforward technique for assessment of nucleic acid replication.

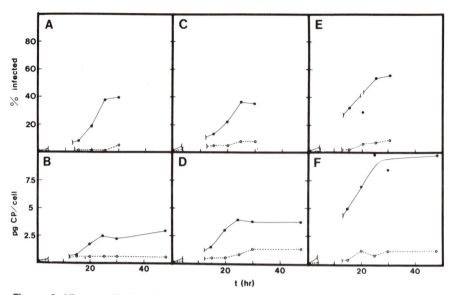

Figure 2. Virus replication in tobacco protoplasts following inoculation with TMV. Mesophyll protoplasts from leaves of either transgenic CP(+) (– – –○– – –) or CP(−) (——●——) plants were inoculated by electroporation with TMV at 10 μg/ml (A,B), 100 μg/ml (C,D), or 1 mg/ml (E,F). Infection was quantified using either an indirect immunofluorescence assay (A,C,E) or a dot immunobinding assay (B,D,F). (Reprinted by permission from ref. 5.)

In addition to the methods of probe generation described in Section 5.3 other options exist if portions of the virus genome have been cloned—*in vitro* transcription of the cloned CP gene has been used in the experiments described here. There are a variety of commercially available vectors that contain well-characterized promoters suitable for use in *in vitro* transcription reactions. The best known of these promoters are from bacteriophages T7 and SP6. With such a vector, strand-specific RNA probes can be produced from cloned DNA inserted in both orientations downstream from the promoter (21). Vectors are even available with different promoters, in opposite orientations, flanking a multiple cloning site. Thus, (+) and (−)strand-specific probes can be made from a single clone by using the appropriate transcription conditions for the desired promoter. For CPMP against TMV, because blockage of both (+) and (−)strand RNA replication is observed in CP(+) cells, attention must be directed to earlier events.

6.3 Blockage of virus uncoating

There are three events in TMV infection prior to RNA replication: virus entry, virus uncoating, and translation of genomic RNA (see *Figure 1*). The need to address uncoating is apparent for two reasons: inoculation of plants or protoplasts with viral RNA (i.e. completely uncoated virus) partially overcomes protection, and early translation of the TMV genome appears to occur concomitant with uncoating.

Metastable, partially stripped TMV rods can be generated by a variety of non-physiological conditions, with the least affected structure being generated by brief exposure of the virus to pH 8.0 (25). In parallel experiments, CP(+) and CP(−) protoplasts and local lesion host plants are inoculated with TMV either held at pH 7.5 or treated briefly at pH 8.0, and infection is assessed. In these experiments, pH 8.0-treated virus overcomes protection to at least the same extent as viral RNA, indicating that the protection is primarily manifest as blockage of a very early stage of uncoating, most likely initiation.

Among the additional lines of support for the results obtained from the above experimental approaches, two are particularly useful to highlight the value of protoplast systems as a novel tool for studying early events of virus infection, and to re-emphasize the value of integrating plant and protoplast experiments in studies of viral resistance at the single cell level. First, non-replicating pseudovirions (which express a marker gene if co-translational disassembly occurs; 26) are apparently unable to uncoat following introduction into CP(+) protoplasts (27). Secondly, although epidermal protoplasts have not been used in the approach described above, the role of the epidermis in protection has been assessed. Transgenic tobacco plants containing a chimeric TMV CP gene that expresses in mesophyll, but not epidermal, cells show less protection against TMV infection than plants which accumulate similar levels of CP in both cell types (7).

7. Concluding remarks

As this chapter has emphasized, novel applications of protoplast/virus systems have become as important as technique development in this field over the past several years. Studies with the TMV/tobacco model system have almost exclusively been presented here because these studies have progressed furthest and because they provide a broad precedent for how to address problems in virus resistance using protoplasts. Recently, protoplast systems have been extended to the study of early events in the replication of spherical viruses (28).

Acknowledgements

Most of the work described here was done in the laboratory of Roger N. Beachy at Washington University in St. Louis where the author was funded by a National Science Foundation Postdoctoral Fellowship. The author thanks Dr Beachy and colleagues in that laboratory for the support and assistance provided during that period.

References

1. Takebe, I. and Otsuki, Y. (1969). *Proc. Natl. Acad. Sci. USA*, **64**, 843.
2. Wood, K. R. (1985). In *Plant cell culture: a practical approach* (ed. R. A. Dixon), pp. 193–214. IRL Press, Oxford.
3. Luciano, C. S., Rhoads, R. E., and Shaw, J. G. (1987). *Plant Sci.*, **51**, 295.
4. Stark, D. M. and Beachy, R. N. (1989). *Bio/Technology*, **7**, 1257.
5. Register III, J. C. and Beachy, R. N. (1988). *Virology*, **166**, 524.
6. Takebe, I. (1984). In *Cell culture and somatic cell genetics of plants* (ed. I. Vasil), Vol. 1, pp. 492–502. Academic Press, Orlando.
7. Clark, W. G., Register III, J. C., Nejidat, A., Eichholtz, D. A., Sanders, P. A., Fraley, R. T., and Beachy, R. N. (1990). *Virology*, **179**, 640.
8. Fannin, F. F. and Shaw, J. G. (1982). *Virology*, **123**, 323.
9. Hibi, T. (1989). *Adv. Virus Res.*, **37**, 329.
10. Murphy, J. F., Jarlfors, U., and Shaw, J. G. (1991). *Phytopathology*, **81**, 371.
11. Paje-Manalo, L. L. and Lommel, S. A. (1989). *Phytopathology*, **79**, 457.
12. Watts, J. W., King, J. M., and Stacey, N. J. (1987). *Virology*, **157**, 40.
13. Okada, K., Nagata, T., and Takebe, I. (1986). *Plant Cell Physiol.*, **27**, 619.
14. Okada, K., Nagata, T., and Takebe, I. (1988). *Plant Cell Rep.*, **7**, 333.
15. Nishiguchi, M., Langridge, W. H. R., Szalay, A. A., and Zaitlin, M. (1986). *Plant Cell Rep.*, **5**, 57.
16. Nishiguchi, M., Sato, T., and Motoyoshi, F. (1988). *Plant Cell Rep.*, **6**, 90.
17. Hibi, T., Kano, H., Sugiura, M., Kazami, T., and Kimura, S. (1986). *J. Gen. Virol.*, **67**, 2037.
18. Saunders, J. A., Smith, C. R., and Kaper, J. M. (1989). *BioTechniques*, **7**, 1124.
19. Hibi, T. and Saito, Y. (1985). *J. Gen. Virol.*, **66**, 1191.
20. Bradford, M. M. (1976). *Anal. Biochem.*, **72**, 248.
21. Sambrook, J., Fritsch, E. F., and Maniatis, T. (ed.) (1989). In *Molecular cloning: a laboratory manual*. Cold Spring Harbor Press, Cold Spring Harbor.
22. Erlich, H. A., Gelfand, D., and Sninsky, J. J. (1991). *Science*, **252**, 1643.
23. Beachy, R. N., Loesch-Fries, S., and Tumer, N. (1990). *Annu. Rev. Phytopathol.*, **28**, 451.
24. Register III, J. C. and Nelson, R. S. (1992). *Seminars in Virol.*, **6**, 441.
25. Wilson, T. M. A. and Shaw, J. G. (1985). *Trends Biochem. Sci.*, **10**, 57.
26. Gallie, D. R., Sleat, D. E., Watts, J. W., Turner, P. C., and Wilson, T. M. A. (1987). *Science*, **236**, 1122.
27. Osbourn, J. K., Watts, J. W., Beachy, R. N., and Wilson, T. M. A. (1989). *Virology*, **172**, 370.
28. Laakso, M. M. and Heaton, L. A. (1994). *Virology*, (in press).

4

Selection of plant cells for desirable characteristics

4A. Inhibitor resistance

ROBERT A. GONZALES

1. Introduction

Selection of cultured cells or protoplasts for inhibitor resistance has yielded a large number of mutants (1, 2). A major reason for selecting mutants is to provide source material for the study of specific biochemical steps leading to a better understanding of metabolic pathways (2–5). Inhibitor resistance is a straightforward method for mutant selection, whereby resistant cells in a large population can be selected by their ability to grow in the presence of the inhibitor while sensitive cells do not.

This section describes the application of standard microbial techniques for the selection of inhibitor resistant mutants from plant suspension cultures. Selections using callus or protoplasts in agar solidified media are readily adapted from the following protocols. For specific examples you are referred to Section 3 of this chapter and refs 5–7.

2. Plant cell culture

There are two major requirements of suspension cell cultures for optimum results in selection experiments.

- **small uniform cell aggregates**—which maximizes contact of inhibitor with all cells in the clump
- **late log phase growth of the inoculum culture**—which maximizes the number of viable cells in the culture

It is not always possible to develop a suspension cell culture with small uniform aggregates. Therefore, it will probably be necessary to pass the inoculum culture through cheesecloth or screens (stainless steel or nylon) to remove the larger clumps. *Figure 1* gives two examples of filter units that can

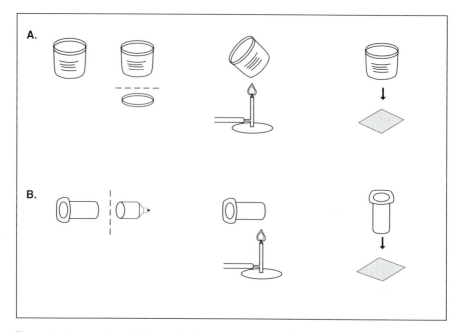

Figure 1. Construction of filter units designed to remove large cell aggregates from plant suspension cultures. A. Specimen container (128 ml). B. Plastic 50 ml syringe. For both units, the bottom is first removed and the cut edge is heated over a flame. The barrel is then pressed on to nylon or stainless steel wire mesh screen material. Large mesh screen material (≥ 1000 μm) works well for sieving cell cultures. The specimen container filter can be placed in a Magenta box and the syringe filter in a 250 ml flask covered with aluminium foil for autoclaving.

be easily constructed for this purpose. The size of the cell aggregates in the inoculum should be as small as practical depending on the ability of the inoculum to grow after filtration.

3. Dose–response curves

The inhibitor concentration used in the selection experiments will depend on the sensitivity of the plant cell line. Therefore, a dose–response curve is first performed for each selective agent of interest. For plant cells, this is usually an end-point growth determination. The range of the inhibitor gradient may be constrained by special considerations such as solubility; however, as broad a range as possible should be chosen. For amino acid analogues, typical gradients range logarithmically from 1 μM to 10 mM (2, 3). *Figure 2* shows a schematic representation of two serial dilution sequences that have been used to prepare gradients in this range. The dilution steps were arbitrarily chosen to facilitate graphic representation of the final results. An example of a dose–

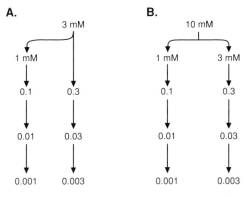

Figure 2. Serial dilution schemes for log dose–response experiments ranging from (A) 1 μM to 3 mM and (B) 1 μM to 10 mM. Concentration gradient steps were arbitrarily chosen to facilitate graphic representation of the data (see *Figure 3*).

response curve generated for a wild-type tobacco cell culture and two tobacco cell lines selected for resistance to ethionine (3) is shown in *Figure 3*.

Protocol 1. Dose–response curve

1. For each cell line to be tested, prepare, in duplicate, a serial dilution of the inhibitor in growth medium. Typically, use 50 ml culture medium in 125 ml flasks.

2. Inoculate each flask with 0.25 g fresh weight[a] of late log phase cells and incubate with shaking at 25 °C for eight to ten days (or until a control culture, in the absence of inhibitor, reaches stationary phase).

3. Harvest each flask by vacuum filtration on Miracloth and determine fresh weight.

4. Average data from duplicate flasks, subtract 0.25 (the original inoculum), and plot on semi-log graph paper. See *Figure 3* for an example. Determine I_{50} (concentration resulting in 50% inhibition) or MIC (minimum concentration resulting in 100% inhibition).

[a] 0.25 g fresh weight per 50 ml culture medium is an arbitrary inoculum empirically determined for tobacco cell cultures; other cell cultures may require a larger inoculum. The important point is that the inoculum size for a given cell culture be standardized throughout the series of experiments.

---

4. Mutant selection

There are two basic strategies for mutant selection from plant cell culture. The first is **single-step selection**, where an inhibitor concentration two to three

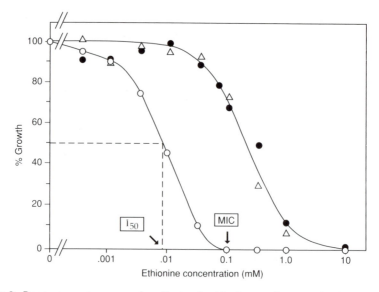

Figure 3. Dose–response curves for effects of ethionine on the growth of sensitive (O) and resistant (● and △) tobacco cell cultures. Cells were harvested ten days after inoculation with 0.25 g fresh weight cells into 50 ml fresh medium containing ethionine. Reprinted by permission from ref. 3.

times the MIC is used. This is the simplest method with clear results and the least likelihood of escapes. The second strategy is **multi-step selection**, where the initial inhibitor concentration is considerably less than the MIC, usually near the I_{50}. Frequent subculture is combined with a gradual increase in inhibitor concentration to allow the faster growing resistant cells to outgrow the sensitive wild-type cells. By gradually increasing the inhibitor levels, it should be possible to eventually obtain vigorously growing cultures at inhibitor levels well above the MIC established for the original unselected culture.

In animal systems, the two selection strategies generally lead to selection of different classes of mutations (8). This is not strictly the case for plant cells (9, 10). However, the choice of strategies will more than likely be determined by the growth characteristics of the cell culture. The single-step method is more effective with fine cultures that exhibit small, reasonably uniform cell aggregates.

Protocol 2. Single-step mutant selection

1. Prepare 1 litre of growth medium containing the inhibitor at two to three times the MIC.

2. Distribute 50 ml aliquots of the selection medium to 20 125 ml culture flasks and autoclave (filter sterilize if the inhibitor is heat labile).

---

3. Inoculate each flask with 0.25 g fresh weight of late log phase plant cells. Incubate with shaking for up to eight weeks.[a]

4. Subculture any flask showing growth to fresh medium containing the inhibitor.

5. Perform a dose–response experiment (*Protocol 1*) with the new cell line to determine resistance level.[b]

---

[a] The length of time for the selection experiment is also arbitrarily determined (i.e. starting from a single cell, a tobacco cell culture will theoretically reach stationary phase by eight weeks).
[b] It is important to maintain the selected culture on medium containing the inhibitor until it has been determined that the resistance phenotype is stable.

Protocol 3. Multi-step mutant selection

1. Prepare culture medium containing a sublethal inhibitor concentration (i.e. MIC > [I] $\geq$ I_{50}).

2. Inoculate a number of flasks[a] with 0.25 g fresh weight of late log phase plant cells per 50 ml of medium.

3. As the cell cultures grow, subculture to fresh medium containing increasingly higher inhibitor levels.[b]

4. When the cell culture continues to grow at inhibitor concentrations above the MIC, perform a dose–response experiment to determine resistance level.[c]

---

[a] The number of flasks inoculated is determined by practical aspects. Each culture will have to be subcultured a substantial number of times during the course of the experiment.
[b] A number of subcultures at the same inhibitor concentration may be necessary before vigorous growth is observed.
[c] It is important to maintain the culture on medium containing inhibitor until the stability of the resistance phenotype has been determined.

---

5. Notes on mutant selection

The mutation rate in a plant cell culture is quite low, and, as a result, the number of inhibitor-resistant cells in the population can be expected to be extremely low. The tobacco suspension cultures, for which the above protocols were developed, contain approximately 3×10^6 cells/g fresh weight. Thus, in the single-step selection protocol, approximately 1.5×10^7 cells are screened in one experiment. In successful selections, one or two resistant cell lines can be selected; a rate of $1–1.3 \times 10^{-7}$. From this, it can be implied that one reason many selection experiments fail is that an insufficient number of cells have been screened. The use of cell suspension cultures is a decided advantage in this respect.

Unfortunately, for one reason or another, the single-step strategy does not yield resistant mutants for all inhibitors. In those cases, it is obligatory to use a multi-step strategy starting with sublethal inhibitor concentrations. However, with the possible exception of gene amplifications (8), which show a higher frequency of occurrence, the same concerns about resistant cell population dynamics should be considered in the experimental design.

Acknowledgement

The author wishes to thank Dr J. M. Widholm, in whose laboratory most of the techniques described in this chapter were developed.

References

1. Maliga, P. (1984). *Annu. Rev. Plant Physiol.*, **35**, 519.
2. Gonzales, R. A. and Widholm, J. M. (1985). In *Primary and secondary metabolism of plant cell culture* (ed. K.-H. Neumann, W. Barz, and E. Reinhard), pp. 337–43. Springer-Verlag, Berlin.
3. Gonzales, R. A., Das, P. K., and Widholm, J. M. (1984). *Plant Physiol.*, **74**, 640.
4. Coruzzi, G. M. (1991). *Plant Sci.*, **74**, 145.
5. Subramanian, M. V., Hung, H.-Y., Dias, J. M., Miner, V. W., Butler, J. H., and Jachetta, J. J. (1990). *Plant Physiol.*, **94**, 239.
6. Hamill, J. D., Ahuja, P. S., Davey, M. R., and Cocking, E. C. (1986). *Plant Cell Rep.*, **5**, 439.
7. Maillot-Vernier, P., Schaller, H., Benveniste, P., and Belliard, G. (1990). *Plant Physiol.*, **93**, 1190.
8. Schimke, R. T. (1984). *Cell*, **37**, 705.
9. Dyer, W. E., Weller, S. C., Bressan, R. A., and Herrmann, K. M. (1988). *Plant Physiol.*, **88**, 661.
10. Goldsbrough, P. B., Hatch, E. M., Huang, B., Kosinski, W. G., Dyer, W. E., Herrmann, K. M., and Weller, S. C. (1990). *Plant Sci.*, **72**, 53.

4B. Cold tolerance

D. C. W. BROWN and J. SINGH

1. Introduction

Plants exhibit a wide-range of responses, when exposed to low temperatures, which have been well-characterized (1, 2). Cold temperature tolerance in plants is most often considered in two broad classes (1–3):

(a) Chilling tolerance based on the ability to withstand exposure to low temperatures above zero.

(b) Freezing tolerance based on the ability to withstand exposure to subzero temperatures.

The degree of chilling tolerance in a plant can be measured in the reversible and irreversible responses of a number of normal plant processes such as seed germination, growth, leaf turgor, photosynthesis, fruit set and yield, or fruit quality. Changes in these parameters reflect a strategy of adjusting biochemical and physiological cell functions to maintain growth under cold conditions. In contrast, freezing tolerance is described in terms of winter hardiness, winter survival, and frost resistance and reflects the more severe, immediate, and lethal consequence of exposure to freezing temperatures and subsequent ice formation within the plant. Dramatic morphological changes, such as leaf abscission and bud formation, concomitant with the cessation of growth, in freezing tolerant plants reflects a strategy for species survival by evasion and avoidance rather than of growth maintenance as observed in chilling tolerant plants.

Plants not only vary in their ability to withstand exposure to cold but plants such as temperate perennials follow a seasonal rhythm of dormancy and freezing tolerance (2). Exposure to low non-freezing temperatures and short photoperiods are triggers which appear to cause a complex range of biochemical and physiological changes associated with an increase in cold tolerance. Alterations in membrane lipid composition, membrane fluidity, isoenzyme pattern, sugar content, soluble proteins, proline, organic acids, abscisic acid, mRNA, and polypeptides have been documented (1–5) and correlated with the acquisition of cold or freezing tolerance. This adaptation is called cold acclimation and represents the transition of plant hardiness from chilling sensitive or tender to a hardy condition. A reversal of the process, deacclimation, occurs when plants are removed from cold temperature. The time for deacclimation is often shorter than for acclimation. Cold acclimation, and deacclimation, can take hours to weeks to accomplish and this has to be taken into consideration when using *in vitro* selection. Furthermore, the process of cold acclimation is considered to be composed of different stages which appear to respond to different stimuli (6):

(a) Stage 1: characterized by a rapid but limited cold acclimation and continued growth in response to photoperiod changes during warm temperature exposure.

(b) Stage 2: characterized by biochemical, physiological, and morphological changes and cessation of growth in response to cold temperatures.

(c) Stage 3: characterized by the development of extreme cold tolerance to the genetic limitation of the plant in response to prolonged continuous exposure to freezing temperatures.

2. Target systems and selection strategies

Several observations influence strongly the strategy used for *in vitro* selection for cold tolerance:

(a) Genetic instability of tissue cultures is a general phenomenon and the resultant variability can be expressed in regenerated plants (7).

(b) Variation in the response to freezing extends to tissue cultured cells, even to the single cell level (8).

(c) Cells acclimate in response to non-freezing temperature or other external stresses such as desiccation (6, 9).

(d) Endogenous levels of some compounds such as proline and abscisic acid appear to increase during acclimation (3, 10).

(e) Application of compounds such as abscisic acid appear to induce cold tolerance, sometimes in the absence of cold exposure (9–11).

The variability of tissue culture-derived plants can be exploited to select low temperature tolerant plants. For example, low temperature flowering variants have been selected from irradiated and non-irradiated cell suspensions of *Chrysanthemum moriflium* grown at low temperatures during regeneration (11, 12). The results suggested that selection was most efficient from slow regenerating (grown at 6 °C), irradiated callus tissue. In our laboratories, we have found that cell suspension cultures of *Medicago sativa* showed an increase in freezing tolerance from −3 °C to −12.5 °C in response to exposure to abscisic acid and a temperature of 2 °C (13). Six lines tested showed different patterns of cold acclimation, with the degree of cold acclimation dependent on the time of cold exposure, the concentration of abscisic acid, and the presence of cytokinin in the medium. Although this type of cell suspension culture (rapid growth, small cell clusters, frequent sub-culture) is desirable for biochemical or physiological study, it exhibits a long cold acclimation period (four weeks) and regeneration of plants from the selected survivors from callus or cell suspension cultures is difficult if not impossible (e.g. 13–15). In contrast, cold acclimation in the presence of abscisic acid, cold, and absence of cytokinin, with a highly embryogenic cell suspension of *Brassica napus* (16), showed optimum cold adaptation occurred within eight days (see *Protocol 1*), and cells surviving exposure (see *Protocol 2*) to −20 °C could undergo plant regeneration after regrowth. Similarly, microspores of *B. napus*, selected at −15 °C after exposure to 2 °C for three weeks and an abscisic acid treatment prior to embryo induction, retained the ability for plant regeneration (17). However, even when regeneration from cold tolerant callus is efficient, recovered plants may not express increased cold tolerance at the plant level, as has been found in studies with carrot (18) and *Nicotiana sylvestris* (19).

Protocol 1. Cold acclimation of *Brassica* cell suspension cultures[a]

1. Establish *Brassica* cell suspension cultures as outlined (20).

2. Prepare culture medium lacking cytokinins. Transfer 49.5 ml of medium into 125 ml Erlenmeyer flasks, cap, and autoclave for 17 min at 121 °C. Reserve half of the medium for washing cells. The use of graduated flasks is recommended for estimation of medium content (see step **8**).

3. Prepare an abscisic acid stock solution by dissolving 100 mg abscisic acid in 2 ml 1.0 M KOH in a 100 ml volumetric flask and then make the volume up to 100 ml with distilled water. The abscisic acid may take several minutes to dissolve even with vigorous shaking. Store at 5 °C in the dark for only a few days if not used immediately.

4. Add the abscisic acid to half of the flasks containing 49.5 ml of sterile medium, by aseptically adding 0.67 ml of abscisic acid stock solution using a 1 ml sterile plastic syringe equipped with a 25 mm 0.22 μm Millex-GS Millipore filter (or equivalent).

5. Wash the cell suspension with medium lacking cytokinin by allowing the cells to settle for 5 min, then carefully pour off the excess medium.

6. Replace the discarded medium with medium devoid of cytokinin and swirl the flask to resuspend the cells.

7. Repeat steps **5** and **6** two times to ensure the removal of residual cytokinin-containing medium.

8. After the third wash, replace with medium containing 5×10^{-5} M (0.67 mg/ 50 ml) filter sterilized abscisic acid to 50 ml by estimation.

9. Place the cell suspensions at 2 °C and grow for six to nine days. This treatment will result in LT_{50} survivals above 50% (see Section 3). In *B. napus*, treatment for less than six days or longer than nine days results in a significant loss of cell freezing tolerance (14).[b]

[a] A similar approach has been shown with alfalfa cell suspension cultures (21).
[b] Alfalfa cell suspensions may take up to four weeks to achieve a similar degree of freezing tolerance (13).

The observation that freezing tolerance extends to the single cell level led to the use of highly regenerable callus cultures for successful plant recovery after cryoselection in wheat (8). Unlike the strategy with cell suspension-derived cells, highly embryogenic callus cultures of *Triticum aestivum* were not cold acclimated prior to selection, and, after being exposed to a controlled freezing rate of between 0.1 °C and 30 °C/minute down to −35 °C (see *Protocol 2*), were immersed in liquid nitrogen. After re-plating for regrowth, a low number of callus pieces were observed to contain regenerable cells and several of these lines showed increased survivability at −12 °C. This increased freezing tolerance was transmitted to the seed progeny of the regenerated plants.

Strategies of indirect selection for cold tolerance based on related biochemical adaptations have also been proposed. Since the accumulation of proline has been observed during cold acclimation (2, 3), it has been suggested that proline over-producers may also show an enhanced level of freezing tolerance. Following this rationale, cell suspension and callus cultures of alfalfa have been selected for growth on the proline analogues *trans*-4-hydroxy-L-proline and L-azetidine-2-carboxylic acid in a search for proline over-producing variants (22). Regenerated plants from selected lines showed high levels of proline accumulation upon hardening and subsequent freezing tests indicated an increased survival level at −14 °C (100% survival in analogue-selected lines verses 25% in controls).

Protocol 2. Slow freezing of cell suspension cells

Equipment

- Programmable cooling bath (to −20 °C)

Method

1. Establish a cell suspension culture (see *Protocol 1* and Chapter 1).
2. Allow cell suspension cultures to settle in the flasks for 5 min.
3. Collect cells by transferring, with a pipette, 0.5 ml samples of cell suspension at the bottom of the culture flask to 16 × 100 mm test-tubes and cap with aluminium foil.[a]
4. Centrifuge at 380 g for 10 min.
5. To minimize the influence of the culture medium, wash the cells three times by carefully pouring off the excess medium after centrifugation and resuspending the cells in 0.5 ml of sterile tap-water. Do not pour off the tap-water at the end of the third wash.
6. Allow the cells contained in the test-tubes to cool at a rate of −2.5 °C/30 min to a lowest temperature of −20 °C, using a Lauda K4/R cooling bath (or equivalent).
7. At −2.5 °C, initiate freezing by adding a small grain of dry ice. This prevents supercooling of the water in the cells (23).
8. At the desired temperature point (e.g. −2.5, −5, −10, −15, −20 °C), remove the test-tube containing the cells and incubate at 4 °C for 20–24 h.
9. Re-plate the cells under the appropriate growth conditions or measure the LT_{50} (see Section 3).

[a] Prepare enough test-tubes to allow for sampling at the desired temperatures (e.g. 0, −2.5, −5, −10, −15, −20 °C). If cells are to be re-plated *in vitro*, then manipulations must be aseptic and test-tubes, caps, slides, cover-slips, and water must be sterile. Also, if cells are to be re-plated, it is convenient to prepare duplicate samples for each temperature point and use one test-tube as an inoculant and the other for an LT_{50} measurement (see Section 3).

3. Assessment of plant survival

Regrowth of cells plated on medium is the best measure of survival after freezing; however, this may require several weeks before results are known and it is difficult to make accurate quantitative measurements of cell freezing tolerance. Assessment of cell survival immediately after recovery can be made using a number of techniques, most of which are based on cellular membrane integrity. Microscope observations of individual cells with respect to their ability to plasmolyse in balanced salt solutions, evidence of protoplasmic streaming (cyclosis), the uptake of vital stains such as neutral red (see *Protocol 3*), or triphenyl tetrazolium chloride, or fluorescent dyes such as fluorescein diacetate (see Chapter 7), the degree of electrolyte leakage, and chlorophyll fluorescence have all been used as an indication of cell survival. It is normal and prudent to use several different methods to assess a population of cells (e.g. protoplasmic streaming plus fluorescent dye plus cell regrowth), as an effective and reliable single freezing test has been difficult to achieve due to environment and genotype interactions, imprecise control of conditions for individual cells, and the possibility of different forms of freezing stress occurring in the isolated cells and tissues (e.g. extracellular versus intracellular freezing and supercooling). A common expression of cellular freezing tolerance is the LT_{50} value for a given population of cells. The LT_{50} is the temperature at which 50% of the cells are killed, and the ranking of genotypes using LT_{50} appears to show general agreement with field performance of cultivars (24).

Protocol 3. Measurement of cell survival with neutral red stain

Equipment and reagents

- Microscope with × 160 magnification
- Neutral red (Sigma Chemical Co.)

Method

1. Prepare a stock solution of neutral red by dissolving 100 mg of neutral red in 200 µl of 1% acetic acid and adding 100 ml of distilled water. Filter through Whatman No.1 filter paper.

2. Remove a small sample of cells from the test-tube using a Pasteur pipette and transfer to a glass microscope slide.

3. Gently wash the cells three times with distilled water to remove the excess culture medium.

4. Add two drops of neutral red solution, one drop of water, and mix gently.

5. Remove any excess solution using a paper towel or filter paper by touching the paper to the edge of the solution.

Protocol 3. *Continued*

6. Gently cover with a cover-slip and view with a microscope under phase contrast at × 160 magnification. Cells taking up the red stain show membrane integrity and are scored as living.

7. Observe at 5 and 30 min.

References

1. Levitt, J. (1972). In *Responses of plants to environmental stresses* (ed. T. T. Kozlowski). Academic Press, New York.
2. Levitt, J. (1980). In *Responses of plants to environmental stresses* (2nd edn). Vol. 1. Chilling, freezing and high temperature stresses (ed. T. T. Kozlowski). Academic Press, New York.
3. Blum, A. (1988). *Plant breeding for stress environments*. CRC Press Inc., Boca Raton, Florida.
4. Thomashow, M. F., Gilmour, S. J., Hajela, R., Hovarth, D., Lin, C., and Guo, W. (1990). In *Horticultural biotechnology* (ed. A. B. Bennett and S. D. O'Neill), pp. 305–14. Wiley-Liss, Inc., New York.
5. Li, P. H. (1989). In *Low temperature and stress physiology in crops* (ed. P. H. Li), pp. 167–76. CRC Press Inc., Boca Raton, Florida.
6. Layne, R. E. C. (1992). *Plant Breed. Rev.*, **10**, 271.
7. Duncan, D. R. and Widholm, J. M. (1986). *Plant Breed. Rev.*, **4**, 153.
8. Kendall, E. J., Qureshi, J. A., Kartha, K. K., Leung, N., Chevrier, N., Caswell, K., and Chen, T. H. H. (1990). *Plant Physiol.*, **94**, 1756.
9. Guy, C. L. (1990). *Annu. Rev. Plant Physiol. Plant Mol. Biol.*, **41**, 187.
10. Chen, H. H., Li, P. H., and Brenner, M. L. (1983). *Plant Physiol.*, **71**, 362.
11. Huitema, J. B. M., Preil, W., Gussenhoven, G. C., and M. Schneidereit, (1989). *Plant Breed.*, **102**, 140.
12. Huitema, J. M. B., Preil, W., and De Jong, J. (1991). *Plant Breed.*, **107**, 135.
13. Orr, W., Singh, J., and Brown, D. C. W. (1985). *Plant Cell Rep.*, **4**, 15.
14. Chen, W.-H., Cockburn, W., and Street, H. E. (1982). *In Plant tissue culture 1982* (ed. A. Fujiwara), pp. 485–6. Japanese Assoc. for Plant Tissue Culture, Maruzen Co., Tokyo.
15. Tumanov, I. I., Butenko, R. G., Ogolevets, I. V., and Smetyuk, V. V. (1977). *Soviet Plant Physiol.*, **24**, 728.
16. Orr, W., Keller, W. A., and Singh, J. (1986). *J. Plant Physiol.*, **126**, 23.
17. Orr, W., Johnson-Flanagan, A. M., Keller, W. A., and Singh, K. J. (1990). *Plant Cell Rep.*, **8**, 579.
18. Templeton-Somers, K. M., Sharp, W. R., and Pfister, R. M. (1981). *Z. Pflanzenphysiol. Bd.*, **103**, 139.
19. Dix, P. J. (1979). In *Low temperature stress in crop plants* (ed. J. M. Lyons, D. Graham, and J. K. Raison), pp. 463–72. Academic Press, New York.
20. Simmonds, D. H., Long, N. E., and Keller, W. A. (1991). *Plant Cell Tiss. Organ Cult.*, **27**, 231.
21. Atanassov, A. and Brown, D. C. W. (1984). *Plant Cell Tiss. Organ Cult.*, **3**, 149.

22. Matheson, S. and Nowak, J. (1991). In *Department of plant science annual report*, pp. 53–5. Nova Scotia School of Agriculture, Truro, NS.
23. Meryman, H. T. and Williams, R. J. (1985). In *Cryopreservation of plant cells and organs* (ed. K. K. Kartha), pp. 13–47. CRC Press Inc., Boca Raton, Florida.
24. Pollock, C. J. and Eagles, C. F. (1988). In *Plants and temperature* (ed. S. P. Long and F. I. Woodward), pp. 157–80. Society for Experimental Biology, The Company of Biologists Limited, Univ. of Cambridge, Cambridge.

4C. *In vitro* selection for salt tolerance

I. WINICOV

1. Introduction

Salt tolerance in crop plants is an increasingly desirable characteristic not only because of the limited water supplies in the world, but also because of salinization of irrigated lands. Since traditional plant breeding methods have been slow to yield substantial improvements in salt tolerance, the alternative approach of utilizing plant cell culture and regeneration of plants from potential cell mutants has received increased attention. An added advantage in generation of salt tolerant plants through tissue culture is that the regenerated plants are clones of the original starting material that differ only in one or a few characteristics, thereby simplifying subsequent testing.

Initial attempts in this area met with very limited success, primarily due to the inability to regenerate vigorous and fertile plants after selection on salt, and the perceived inability to obtain plants with changes in an apparent multigenic trait that could be transmitted through seed to successive generations of plants.

Although we still do not understand the mechanism(s) by which plants can acquire incremental improvements in their salt tolerance, recent reports show that fertile plants can be regenerated from cells after selection on salt in culture and that the salt tolerance trait can be transferred through seed. These results indicate that the salt tolerance trait expressed in cell culture can contribute to increased salt tolerance of the whole plant. Thus, the generation of salt tolerant plants through tissue culture becomes a viable alternative to classical plant breeding in improving the salt tolerance of crop plants. Successful application of this method, however, requires that the salt tolerance trait selected in culture satisfies the following criteria:

- it is stable in tissue culture
- it allows efficient regeneration of plants
- it confers improved salt tolerance to plants regenerated from such a culture
- it is transferred through seed to progeny

This list defines a mutation that is selected through somaclonal variation, which should occur at frequencies characteristic for mutations and which eventually should be amenable to identification by molecular means. This chapter examines the relationship of somaclonal variation and salt tolerance and endeavours to develop some guide-lines for the successful use of this method in generating salt tolerant plants.

2. Cellular salt tolerance

2.1. Somaclonal variation and salt

Spontaneous somaclonal cell variants, tolerant to NaCl, have been selected on media containing NaCl in several plant species (1, 2). Tolerance to NaCl has also been achieved by selection on a combination of salts (3–5). The frequency of spontaneous variants in tissue culture is influenced by the increase in frequency of chromosomal loss or rearrangement (6) and it has been reported that growth on NaCl induces an even higher level of somaclonal variation in culture (7). In alfalfa such variation in culture may be enhanced by endogenous transposon activation (8). However, while the increased tendency for somaclonal variation in cultures grown on salt may enhance the selection process through increasing the number of variants, it may also work against achieving vigorous whole plant regeneration by introducing unrelated mutations that affect whole plant vigour and fertility.

2.2 Isolation of stable variants

The ability to select salt tolerant somaclonal variants from cultures that are normally salt sensitive implies changes in regulation of the genes that are already present in the genomic make-up of the original plant cell (9, 10). This is demonstrated by the increase in expression of some genes and decrease in others as a result of selection for salt tolerance in tissue culture (11, 12). Selection for stable salt tolerance in culture helps to select against lines with epigenetic changes. Extra effort at this stage is likely to reduce the number of plants that turn up as 'not-tolerant' after screening of the regenerated plants.

3. Characteristics of salt tolerance at the regenerated plant level

The incremental improvement in salt tolerance that is obtained by selection in tissue culture may occur through changes in regulation or over-expression of genes for a number of physiological pathways that are either highly susceptible to salt stress or contribute to molecular mechanisms that protect the plant from such a stress. We have identified one such pathway in our alfalfa cell cultures (10, 13), but others are equally likely to provide incremental improvements in tolerance. The regenerated salt tolerant alfalfa appear morphologically

indistinguishable from their salt sensitive parents and the growth characteristics of the F-2 seedlings are comparable to those of the parent seedlings (14).

Increased vigour has been observed in several plants selected through
survival on salt *in vitro* followed by immediate regeneration (15). In some
cases, such selections have involved large populations of seed-initiated callus
or direct explants (16–19). The specific role of the increased vigour in salt
tolerance is not understood, but may add another dimension to the regeneration method, since some of the non-selected regenerated plants also showed
improved salt tolerance when compared to control plants.

4. Generation of salt tolerant alfalfa plants through tissue culture

The protocols described here have been successfully used for alfalfa (14) and
have been specifically designed to minimize the time in culture in order to
achieve tolerance with a minimum of undesirable characteristics.

You will need access to the following equipment and facilities:

* sterile tissue culture facilities
* lighted, temperature controlled incubator
* growth chamber or greenhouse

4.1 Tissue culture

4.1.1 Initiation

To obtain young tissue cultures for the salt selection protocol, initiate new
cultures from suitable plant material, as described in *Protocols 1* and *2*. We
have used both immature alfalfa ovaries and alfalfa cotyledons from seed
germinated on Schenk and Hildebrandt (20) medium (SH) without hormones
(see Chapter 1).

Protocol 1. Initiation of callus from immature alfalfa ovaries

1. Harvest buds from alfalfa

2. Sterilize buds by dipping successively in the following solutions:
   * 95% ethanol, for 90 sec
   * 20% chlorox bleach with about 0.5% soap, for 90 sec
   * sterile distilled water, wash twice
   * 95% ethanol, for 90 sec

3. Transfer aseptically to container with sterile distilled water.

4. Dissect ovaries from each bud aseptically and place several (five or six)
   on a sterile culture plate (Falcon, Optilux No. 1005) containing 50 ml of

---

Protocol 1. *Continued*

    callus proliferation (CP) medium (modified SH medium with 2 mg/litre 2,4–D, 2 mg/litre kinetin, solidified with 6 g/litre agar (Difco, BiTek™)).

5. Seal each dish with Parafilm to maintain moisture and incubate at 24 ± 1 °C under continuous light (approximately 30 $\mu E \cdot m^{-2} \cdot s^{-1}$). Callus formation should become apparent in about a week.

Protocol 2. Initiation of callus from cotyledons

1. Scarify dormant alfalfa seed by rubbing gently with fine sandpaper.
2. Sterilize the seed by successively adding the following solutions, aspirating each after the indicated time period:
   - 95% ethanol, swirl the seeds for about 10 sec
   - 20% chlorox with 0.1% Tween 40—swirl the seed in this solution, cover, and allow to sit for 20 min, aspirate the chlorox–Tween solution
   - 20 ml sterile distilled water, wash the seeds three times
3. Put seeds on sterile solidified SH medium without hormones.
4. Set in light until sprouted and the cotyledons are developed (three to four days).
5. Cut off the cotyledon first leaf aseptically and place on CP medium with hormones (see *Protocol 1*, step 4). Continue as in *Protocol 1*.

4.1.2 Growth requirements

Allow the newly initiated callus to grow for four weeks on CP medium as described in *Protocol 1*, step 4. At this point select calli that show rapid growth for continued propagation and selection for salt tolerance. The calli are propagated by aseptically transferring about 200–250 mg callus to fresh CP medium with five callus pieces spaced on each plate. The parent salt sensitive line is maintained throughout the experimental period by division and transfer every four weeks, while selection for salt tolerance can begin as soon as sufficient callus material has been generated by growth on the CP medium. Before proceeding with the selection process:

(a) Estimate the number of cells per given weight or size of callus. Gently disrupt the callus with a glass pestle in a known volume of buffer and count the cells in a haemocytometer.

(b) Amplify the salt sensitive callus to yield enough cells for selection of variants at frequencies of 10^6 to 10^7.

(c) Estimate experimentally the amount of salt that is lethal to more than 99% of your cell population.

4.2 Selection for cellular salt tolerance

The selection process is critical if healthy and fertile plants are to be regenerated subsequently. The primary considerations in our experience with alfalfa are:

(a) The cells should be in culture only a limited time before the selection process, preferably no more than three months.

(b) The selection should be a single-step selection with a salt concentration that is lethal to more than 99% of the cells.

(c) Salt selection of callus is likely to introduce fewer genetic rearrangements than selection in suspension culture.

4.2.1 Selection of tolerant cells

Select tolerant cells with the appropriate salt concentration (1% or 171 mM NaCl for alfalfa) as shown in *Protocol 3*.

Protocol 3. Selection of salt tolerant cells

1. Prepare plates containing 50 ml solidified CP medium with 1% NaCl.

2. Plate the cultured cells as callus pieces no larger than 5 mm in diameter on the salt containing CP medium, plating five to six pieces per plate.

3. Seal with Parafilm and incubate in continuous light as described in *Protocol 1*, step 5 for four to six weeks.

4. After incubation on salt, most of the calli should be discoloured and dead. However, you will notice occasional foci of live cells. Select those that appear potentially embryogenic and transfer to fresh salt containing CP medium, taking care to avoid carrying along most of the dead cells.

5. Grow the selected calli for two additional subcultures on CP plus salt, to increase the mass of each tolerant culture.

6. Select isolated cell lines that grow rapidly on the previously lethal salt concentration and proceed simultaneously with testing of the salt tolerant line (*Protocol 4*) and regeneration (*Protocol 5*).

4.2.2 Characterization of selected lines

Selected salt tolerant cell lines should always grow equally well or better on salt containing CP medium than on control CP medium. Cell lines that are stably salt tolerant will not lose their ability to grow on a high level of salt after a two month subculture on control medium. Salt tolerant lines which arose through epigenetic changes would be less likely to maintain their tolerance in the absence of selective pressure by NaCl. Most of the alfalfa lines selected to date have shown stable salt tolerance in cell culture.

Protocol 4. Testing of cellular salt tolerance

1. Plate at least three equal size callus pieces of each selected line on CP and CP plus salt medium. Incubate for four weeks in light as in *Protocol 1*, step 5. Note visually if comparable growth can be obtained on CP and CP plus salt medium.

2. Transfer each culture on the same medium as plated in step **1**. Incubate for an additional four weeks and again note growth on each medium.

3. Subdivide each callus and re-plate each on CP and CP plus salt medium. You should have four sets of calli from each line:
   - calli from CP on CP and CP plus salt
   - calli from CP plus salt on CP plus salt and CP

   Incubate for four weeks as above and again note the growth characteristics. Quantitate the results by weighing the calli.

4.3 Regeneration of alfalfa plants from salt tolerant cell culture

Regeneration should be initiated from the selected salt tolerant cell lines as soon as it becomes apparent that a cell line has been isolated that grows equally as well on salt as in the absence of salt and sufficient amount of callus becomes available. We found that callus maintained for eight months on salt, yielded inferior plants in both vigour and ability to flower (14). Alfalfa regeneration from cell culture occurs predominantly through embryogenesis, and regeneration of Regen S-derived lines is very straightforward. Usually protocols that give the highest levels of regeneration for the plant cells selected should be used, while in all cases keeping the time in culture at the minimum. Include non-selected callus from one or more lines to regenerate control plants for subsequent growth comparisons in plant tolerance tests.

Protocol 5. Alfalfa regeneration from salt tolerant cell lines

1. Transfer 1 cm size callus pieces from salt tolerant cells grown on CP plus salt to plates containing 50 ml modified SH medium containing no hormones, solidified with 6 g/litre agar.[a] Incubate in continuous light as in *Protocol 1* and after two to four weeks shoots and roots start to appear.

2. Transfer the plantlets after four weeks to a taller container (Erlenmeyer flask, transparent plastic container, etc.) containing the same medium for continued development.

---

3. When plantlets reach 3–4 cm size, transfer them from the agar directly into sterile distilled water, gently remove agar from around the roots and plant in Jiffy peat containers. Enclose the plants in a sealed transparent plastic container and continue to grow in the incubator for an additional two to four weeks. Some plantlets at this point do not thrive and should be discarded.

4. When plants appear established and growing in peat, harden them to the atmosphere by placing them for three or four days on a mist bench, or, alternatively, increasingly remove the container cap for several days. Transplant to soil in the greenhouse.

[a] The regeneration protocol described here may need to be supplemented by additional steps, such as a transient exposure to high levels of 2,4-D, increased sources of nitrogen, etc. as described in Chapters 1 and 5.

4.4 Testing regenerated plants for improved salt tolerance

To test the regenerated plants for improved salt tolerance, root replicate cuttings of the regenerated plants, both selected and non-selected, as well as the parent plant from which the salt sensitive callus was derived. Measure both survival and growth of the plants with increased salt exposure.

Protocol 6. Greenhouse testing of salt tolerance of regenerated plants

1. Establish at least 15 equal rooted cuttings of each selected plant and each control plant in Conetainers™ in perlite which is watered daily, to flush out accumulated salts, followed by one quarter strength Hoaglands solution (21).

2. After four weeks cut back the shoots and divide the plants into three groups with at least five replicates of each plant in each group.

3. Continue watering daily for a minimum of four weeks as in step **1**, but include NaCl in the quarter strength Hoaglands solution. The treatments will be:
   - group I (control), no salt
   - group II, 0.5% NaCl
   - group III, 1% NaCl

Tolerance can be expressed by the number of survivors per number of replicate plants in each group, and plant growth by harvesting and weighing the shoots of each plant after four weeks. Compare growth characteristics between the selected and non-selected regenerated plants as well as the parent salt sensitive plant. Selfed seed can be germinated and the tolerance of

F-2 seedlings tested in a similar manner. The ultimate proof of improved salt tolerance lies in field testing.

References

1. Stavarek, S. J. and Rains, D. W. (1984). In *Salinity tolerance in plants: strategies for crop improvement* (ed. R. C. Staples and G. H. Toenniessen), pp. 321–34. John Wiley and Sons, New York.
2. Tal, M. (1984). In *Salinity tolerance in plants: strategies for crop improvement* (ed. R. C. Staples and G. H. Toenniessen), pp. 461–88. John Wiley, New York.
3. Yano, S., Ogawa, M., and Yamada, Y. (1982). *Proc. 5th Int. Cong. Plant Tiss. Cell Cult.*, 495.
4. McHughen, A. and Swartz, M. (1984). *J. Plant Physiol.*, **117**, 109.
5. Freytag, A. H., Wrather, J. A., and Erichsen, A. W. (1990). *Plant Cell Rep.*, **8**, 647.
6. McCoy, T. J., Phillips, R. L., and Rines, H. W. (1982). *Can. J. Genet. Cytol.*, **24**, 37.
7. McCoy, T. J. (1987). *Plant Cell Rep.*, **6**, 417.
8. Groose, R. W. and Bingham, E. T. (1986). *Plant Cell Rep.*, **5**, 104.
9. Winicov, I. (1990). In *Progress in plant cellular and molecular biology* (ed. H. J. J. Nijkamp, L. H. W. van der Plas, and J. van Aartijk), pp. 142–7. Kluwer, Dordrecht, The Netherlands.
10. Winicov, I. and Button, J. D. (1991). *Planta*, **183**, 478.
11. Singh, N. K., Handa, A. K., Hasegawa, P. M., and Bressan, R. A. (1985). *Plant Physiol.*, **79**, 126.
12. Winicov, I., Waterborg, J. H., Harrington, R. E., and McCoy, T. J. (1989). *Plant Cell Rep.*, **8**, 6.
13. Wincov, I. and Seemann, J. R. (1990). *Plant Cell Physiol.*, **31**, 1155.
14. Winicov, I. (1991). *Plant Cell Rep.*, **10**, 561.
15. McHughen, A. (1987). *Theor. Appl. Genet.*, **74**, 722.
16. Vajrabhaya, M., Thanapaisal, T., and Vajrabhaya, T. (1989). *Plant Cell Rep.*, **8**, 411.
17. Jain, R. K., Jain, S., Nainawatee, H. S., and Chowdhury, J. B. (1990). *Euphytica*, **48**, 141.
18. Kirti, P. B., Hadi, S., Kumar, P. A., and Chopra, V. L. (1991). *Theor. Appl. Genet.*, **83**, 233.
19. Ibrahim, K. M., Collins, J. C., and Collin, H. A. (1992). *Plant Cell Tiss. Organ Cult.*, **28**, 139.
20. Schenk, R. U. and Hildebrandt, A. C. (1972). *Can. J. Bot.*, **50**, 199.
21. Hoagland, D. R. and Arnon, D. I. (1938). *Univ. of Calif. Berkeley Agric. Exp. Sta. Circ.*, **347**, 39.

4D. *In vitro* selection for disease/toxin resistance

JAN BRAZOLOT, KANG FU YU, and K. PETER PAULS

1. Introduction

Plant tissue cultures treated with toxins from disease-causing organisms are useful for examining plant–pathogen interactions and improving the disease resistance of crop plants. In particular, many of the molecular details of the hypersensitive reaction have been obtained from studies of crude or purified fungal filtrate preparations and plant tissue cultures (1). Also, pathogen preparations have been shown to contain specific or general toxins that have been used to select disease-resistant plant material (2).

In vitro selection for disease resistance is advantageous for several reasons including that:

(a) Experimental units can be maintained on defined media and in rigorously controlled environments allowing for selection of small increments in disease resistance.

(b) Cultured cells can be uniformly exposed to the selective agents, thus reducing the incidence of escapes.

(c) Culture systems maintained in small spaces can potentially replace expensive greenhouse or field testing facilities.

(d) The disease-causing agent remains confined to the laboratory.

This chapter describes protocols for:

- obtaining a toxic fungal culture filtrate
- measuring the effect of the toxic fungal culture filtrate on callus cultures
- using the fungal toxin to select resistant plant material
- regenerating plants from resistant cultures
- evaluating the disease resistance of the regenerated plants

The protocols are based on methods that were developed for studying Verticillium wilt in alfalfa but the general principles should be applicable to a wide variety of pathogen–plant interactions.

2. Toxic preparations from disease-causing organisms

In an *in vitro* selection experiment, plant tissue cultures are exposed to toxic preparations obtained from phytopathogens. These preparations can be

87

subcellular fractions from the disease-causing organism or can be compounds released into media used to grow the pathogen (as described in *Protocol 1*). Both crude and purified preparations have been used. A variety of compounds have been identified as toxic agents in fungal and bacterial culture filtrates including: peptides, carbohydrates, glycopeptides, lipids, and a variety of small molecules (1–4). It is not unusual for fungal culture filtrates to contain a mixture of toxic substances including heat-stable and heat-labile components. For example, autoclaving the filtrate from *Verticillium* cultures reduced but did not eliminate its toxicity to alfalfa cultures (5). In the majority of the cases the mode of action of the toxic compounds is unknown, but recent studies suggest that their toxic effects may be mediated by free radical mechanisms (1, 4).

Protocol 1. Preparation of fungal culture filtrate (see *Figure 1*)

Equipment and reagents

- Potato dextrose agar (BBL, Becton Dickinson Co., Cockeysville, MD, USA)
- B5 media (6)
- Petri dishes (15 mm × 100 mm)
- Erlenmeyer flask (250 ml)
- Growth cabinet (temperature and light regulated)
- Platform shaker
- Whatman No. 1 filter paper
- Centrifuge capable of spinning 50 ml centrifuge tubes at 10 000 g
- 50 ml centrifuge tubes
- 0.22 µm filter sterilization unit (Nalgene Co., Rochester, NY, USA)

Method

1. Maintain *Verticillium alboatrum* by transferring a 5 mm square of mycelium-containing potato dextrose agar (PDA) from a 30-day-old culture to a new PDA dish every month and incubating the culture at 22 °C in low light (0.5 $\mu E \cdot m^{-2} \cdot s^{-1}$).

2. Inoculate 50 ml of liquid B5 medium (without plant hormones) in a 250 ml Erlenmeyer flask with 10–15 pieces of PDA containing mycelium cut from a two-week-old *Verticillium* culture dish.

3. Incubate the liquid cultures at 22 °C in low light (0.5 $\mu E \cdot m^{-2} \cdot s^{-1}$) with agitation (90 r.p.m.).

4. Subculture after two weeks by adding 50 ml of fresh medium to the culture and dividing it between two flasks (by six weeks the culture should consist of a uniform suspension of mycelial balls approximately 2 mm in diameter).

5. Prepare the cell-free filtrate from cultures that have been subcultured for a minimum of six weeks and a maximum of four months.

6. Collect the filtrate from cultures two weeks after their last subculture by filtering them by gravity through a Whatman No.1 filter.

7. Centrifuge the filtrate at 10 000 *g* for 20 min and sterilize the supernatant by passing it through a 0.22 μm filter unit attached to a vacuum source.

8. Use the fungal culture filtrate immediately at various dilutions[a] or freeze it at −20°C.[b]

[a] A common method of standardizing the concentration of the filtrate preparation is to perform an anthrone test for total glucose equivalents (7).
[b] The *V. alboatrum* culture filtrate also retains its toxicity after freeze-drying.

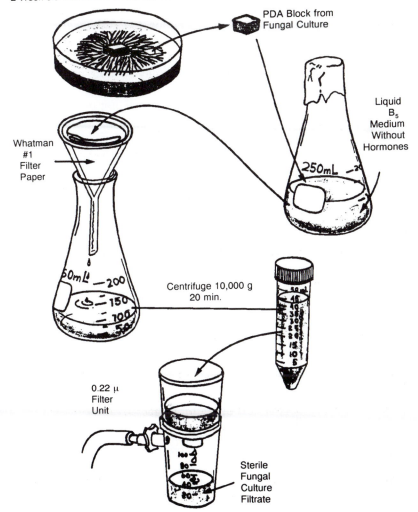

2-Week-old *Verticillium alboatrum* Culture

PDA Block from Fungal Culture

Liquid B$_5$ Medium Without Hormones

Whatman #1 Filter Paper

250mL

50mL — 200 — 150 — 100 — 50

Centrifuge 10,000 g 20 min.

0.22 μ Filter Unit

Sterile Fungal Culture Filtrate

Figure 1. *V. alboatrum* fungal culture filtrate preparation.

3. Plant tissue cultures for *in vitro* selection

In vitro selection can be used with a population of cells that have different genetic compositions, such as microspore or anther cultures, but most often it is used with cultures derived from somatic tissues that are, to a large extent, genetically identical. In the latter case selection acts to isolate rare spontaneous mutants that show resistance to the toxic components in the culture filtrate; these mutants are usually called somaclonal variants (8). Treatments with ionizing radiation or chemicals that modify DNA structure or inhibit cell division are sometimes used to increase the mutation frequency in plant cultures used for selection (9).

Both callus and suspension cultures are useful for studies of plant pathogen interactions (see *Protocol 2* and *Protocol 3*). Callus cultures allow the effects of many different concentrations and modifications of the filtrate on plant cells to be rapidly screened. Suspension cultures may be more appropriate if the objective of the experiment is to isolate resistant plant material because the smaller cell clusters that are found in these cultures are more uniformly exposed to the toxic agent.

Protocol 2. Measurement of the effect of the fungal culture filtrate on alfalfa callus growth

Equipment and reagents

- Alfalfa petioles from young leaves
- Laminar flow bench
- Sodium hypochlorite 5.6% (commercial bleach solution 100%)
- Sterilized distilled water
- Petri dishes (15 × 100 mm)
- B5h media (10) containing B5 salts plus 1 mg/litre 2,4-D and 0.2 mg/litre kinetin (add 8% agar for solid media)
- Petri plates containing solid B5h plus various percentages of *Verticillium* culture filtrate (see *Protocol 1*)

- Parafilm (American Can Co., Greenwich, Conn., USA)
- Tissue culture incubator with lights
- Waring blender and sterile mini-sample blender heads (Fisher Scientific, Unionville, Ontario)
- Autoclaved 7 cm Whatman No. 1 filter paper disks (may be necessary to repeat autoclaving to ensure sterility)
- Wide-mouthed 5 ml pipettes
- Balance

Method

1. Surface sterilize the alfalfa petioles by placing them into 5.6% sodium hypochlorite in a laminar flow bench for 60 sec.
2. Rinse the petioles three times with sterilized water.
3. Remove the bleached ends and cut the petioles into 1 cm pieces.
4. Transfer the pieces to 15 × 100 mm dishes containing B5h, and seal the dishes with Parafilm to prevent desiccation and contamination.
5. Incubate the cultures in a growth cabinet set at 25 °C and a 16 h light, 8 h dark photoperiod (90 $\mu E \cdot m^{-2} \cdot s^{-1}$) for three to four weeks to allow callus to grow from the petiole pieces.

6. Add weighed amounts of callus to liquid B5h in the mini-sampler blender head to give a ratio of 1 g callus/4 ml medium.

7. Homogenize the callus with three 3 sec pulses of the Waring blender set to low speed.[a]

8. Add a 1 ml aliquot of well-mixed callus homogenate, with a 5 ml wide-mouthed pipette, on to a sterile Whatman No. 1 filter placed on solid B5h medium (with or without the fungal filtrate) in 15 × 100 mm Petri dishes.

9. Spread the homogenate evenly over the top filter paper by gently swirling the dish.

10. Two hours after adding the homogenate weigh each culture without the lid.

11. Gently lift the filter paper from the Petri dish with two sterile forceps and record the weight of the dish plus the medium.

12. Determine the weight of the callus by subtracting the weight determined in step **11** from the average weight for the filter paper.

13. Seal the dishes with Parafilm and incubate under the same conditions described in step **5**.

14. Repeat the weighing procedure every four days.

15. Express the callus growth in the dishes containing the culture filtrate as a percentage of the growth in the control dishes.

[a] The homogenization allows the cells to be spread uniformly across the filter paper.

Protocol 3. Measurement of the effect of the fungal culture filtrate on alfalfa suspension culture growth and viability (see *Figure 2*)

Equipment and reagents

- Three- to four-week-old alfalfa callus (see *Protocol 2*)
- Erlenmeyer flasks
- B5g media (10)
- 10 ml graduated centrifuge tube
- Bench-top centrifuge
- Fluorescein diacetate (FDA)
- Microcentrifuge tube 1.5 ml
- Microcentrifuge
- Freezer
- Spectrofluorometer
- Polytron homogenizer

Method

1. Add approximately five petiole calli (see *Protocol 2*, step 5) to a 125 ml Erlenmeyer flask containing 25 ml B5g media.

2. Agitate flasks at 110 r.p.m. for seven days at 25 °C under 16 h light, 8 h dark photoperiod (90 $\mu E \cdot m^{-2} \cdot s^{-1}$).

91

Protocol 3. *Continued*

3. Transfer 20 ml of a seven-day-old suspension culture to a 250 ml flask containing 12 ml of B5g plus 8 ml B5 for a cycle one control culture; or to a flask containing 12 ml of B5g plus 8 ml of the fungal culture filtrate (see *Protocol 1*) for a cycle one treatment culture with 20% (v/v) fungal culture filtrate.

4. Subculture control and treatment cultures at 12 day intervals by adding 20 ml of the previous cycle culture to 20 ml of fresh B5g media containing 20% B5 or 20% filtrate (16 ml B5g plus 4 ml B5 or 4 ml filtrate), respectively.

5. On day 0, 4, 8, and 12 of each cycle transfer 1 ml aliquots from the filtrate-treated and control cultures to graduated test-tubes and add 10 μl of FDA solution (5 mg/ml of acetone) to the samples.

6. Vortex the samples and allow them to stand at room temperature for 15 min; vortex again, and allow the samples to stand for another 10 min.

7. Adjust the volume of the samples to 4 ml with distilled water and centrifuge for 5 minutes at 350 *g* for 5 min.

8. Record the packed cell volumes (PCVs) of the samples and discard the supernatants.

9. Resuspend the cells in 2 ml of distilled water, centrifuge the samples at 350 *g* for 2 min, and discard the wash.

10. Resuspend the cells in 1 ml of distilled water and store the samples at −70°C until all the samples are collected.

11. Bring the volume of the samples up to 3.5 ml with water and homogenize them with a Brinkmann polytron at high speed for 20 sec.

12. Centrifuge the samples at 1400 *g* for 20 min.

13. Dilute the supernatant 1 in 1000 or 1 in 10 000 with distilled water.

14. Measure the fluorescence of the diluted sample at 516 nm with an excitation of 488 nm.

15. Calculate the fluorescence intensity per volume of packed cells to determine relative viabilities of the cultures.

4. Selection strategies

Short-term exposures (several hours) of plant cell cultures to fungal culture filtrates are usually sufficient to induce the variety of biochemical responses observed in the hypersensitive response (1). However, long-term exposures (several weeks) are usually required to select resistant plant material. We have found that repeated subculturing on to fresh media and fresh *Verticillium* culture filtrate results in a successive increase in resistance to the toxin in

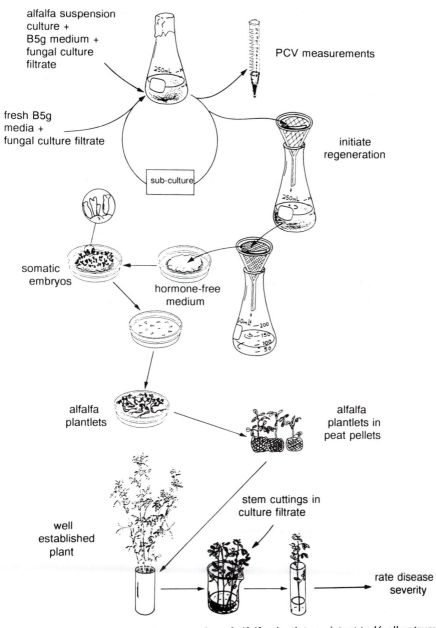

Figure 2. *In vitro* selection and regeneration of alfalfa plantlets resistant to *V. alboatrum* fungal culture filtrate.

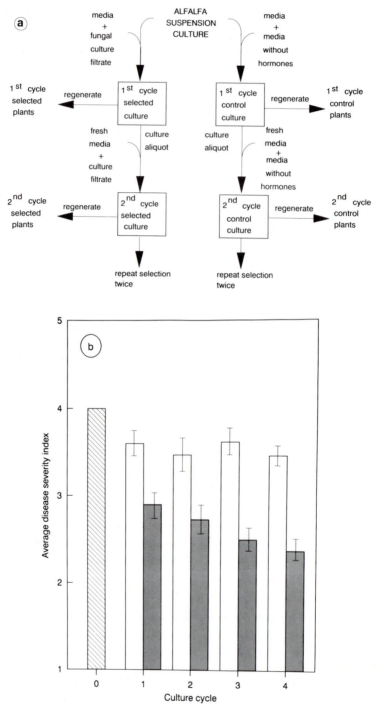

Figure 3. Recurrent *in vitro* selection of alfalfa with *V. alboatrum* fungal culture filtrate. (a) Selection scheme for alfalfa cultures exposed to the *V. alboatrum* fungal culture filtrate for four cycles. (b) Disease severity indices (1 = resistant, 5 = susceptible) for populations of alfalfa plants regenerated from cell cultures exposed to four cycles of 20% (v/v) B5 (☐) or 20% (v/v) *V. alboatrum* fungal culture filtrate (▨); (◩) is a non-cultured control.

the plant material from cycle to cycle (5) and the resistance of plants regenerated from the cultures to the disease (see *Figure 3*).

The degree of selection pressure imposed by the toxic pathogen preparation is also important to control. Usually, sublethal levels of the toxin are used, which inhibit the growth of the plant cell culture to levels that are 1–50% of untreated cultures (2).

5. Regenerating and testing plant material for disease resistance

The applicability of *in vitro* selection to plant improvement is dependent on the expression of totipotency by the undifferentiated selected cells. However, a common limitation of long-term maintenance of plant cultures is the loss of the regenerative capacity of the material over time in culture. If the objective of the *in vitro* selection procedure is to obtain resistant plants then a scheme with alternating cycles of selection and regeneration may be the best approach.

An important step in the use of *in vitro* selection for plant improvement is to test the relationship between resistance of the plant cultures to the culture filtrate and resistance of the plants regenerated from the cultures to the fungus. This is necessary because resistance to a fungal toxin at the cellular level may or may not be related to whole plant resistance to the fungus. Finally, the heritability of the disease resistance obtained by *in vitro* selection must be established if plants that are obtained by this procedure are to be used in a plant breeding programme.

Protocol 5. Regeneration of plants from filtrate-treated and control cultures via somatic embryogenesis (see *Figure 2*)

Equipment and reagents

- 12-day-old alfalfa cultures
- Nylon screens (500, 224, and 63 μm)
- Petri dishes containing BiO2y media (10)
- Peat pellets
- Plastic trays and transparent covers
- Pots
- Planting mix (Turface)

Method

1. Filter 10 ml of the filtrate-treated and control cultures, separately, through the 500 μm mesh and discard the clumps that are collected.

95

Protocol 5. *Continued*

2. Collect the proembryogenic masses from the filtered cultures by passing them through a 224 μm mesh.

3. Transfer the proembryogenic cells on to a fine nylon screen (63 μm mesh) and place the screens plus the cells into a Petri dish containing solid BiO2y medium.

4. Incubate the dishes in a growth cabinet set at 25 °C and a 16 h light, 8 h dark photoperiod (90 $\mu E \cdot m^{-2} \cdot s^{-1}$) for three to four weeks to allow the embryos to develop.

5. Transfer approximately 20 well-formed embryos to dishes with fresh BiO2y media.

6. Incubate the plates at 4 °C in the dark for two weeks and then transfer them back to a growth cabinet set at 25 °C and a 16 h light, 8 h dark photoperiod (90 $\mu E \cdot m^{-2} \cdot s^{-1}$) for three to four weeks to allow the embryos to develop shoots and roots.

7. Transfer plantlets formed from the embryos to moist peat pellets and place into trays covered with transparent plastic lids.

8. Maintain plantlets in a growth room or glass house (23/16 °C and 16/8 h light (400–500 $\mu E \cdot m^{-2} \cdot s^{-1}$)/dark) with the lids until plantlets are acclimated to growth conditions out of culture.

9. When plants are approximately 4–6 cm high transplant them to plastic pots containing a planting mix.

Protocol 6. Determination of disease resistance of regenerated plants

Equipment and reagents

- *Verticillium* culture filtrate
- Test-tubes
- Alfalfa plantlets regenerated from culture
- Sterilized beaker

Method

1. Prepare four stem cuttings per plant from the tops of actively growing plants so that each cutting consists of the apex plus three to four nodes.

2. Place the cuttings into a beaker containing 50% *Verticillium* culture filtrate plus 50% sterilized water for 24 h in a growth cabinet set to 18 °C with 16 h illumination (400–500 $\mu E \cdot m^{-2} \cdot s^{-1}$).

3. Transfer the cuttings into individual test-tubes containing sterilized water and seal the space between the stem and the top of the test-tube with Parafilm.

4. Maintain the cuttings in a growth cabinet set at 18 °C with 16 h illumination (400–500 $\mu E \cdot m^{-2} \cdot s^{-1}$) for seven days and rank the disease severity according to a scale from 1–5 described by Ireland and Leath (11), where:

1 = green leaves
2 = general chlorosis of the leaflets
3 = leaf necrosis with or without chlorosis and leaf curl on less than 50% of the leaves
4 = leaf necrosis and curl on more than 50% of the leaves
5 = wilted stem with chlorosis, necrosis, and leaf curl

The cuttings with ratings of 1 or 2 are considered to be resistant, whereas those with ratings of 3–5 are considered to be susceptible. The disease severity index of each plant is calculated from the mean of the four cuttings.

Acknowledgements

We gratefully acknowledge the assistance of Jennifer Kingswell with the preparation of the manuscript and figures. The work reported from the authors' laboratory was supported by the Ontario Ministry for Food and Agriculture, Agriculture Canada, and the Natural Sciences and Engineering Research Council of Canada.

References

1. Dixon, R. A. and Lamb, C. J. (1990). *Annu. Rev. Plant Physiol. Plant Mol. Biol.*, **41**, 339.
2. Jones, P. W. (1990). In *Plant cell line selection* (ed. P. J. Dix), pp. 113–49. VCH, Weinheim.
3. Ryan, C. A. and Farmer, E. E. (1991). *Annu. Rev. Plant Physiol. Plant Mol. Biol.*, **42**, 651.
4. Gross, D. C. (1991). *Annu. Rev. Phytopathol.*, **29**, 247.
5. Frame, B., Yu, K.-F., Christie, B. R., and Pauls, K. P. (1991). *Physiol. Mol. Plant Pathol.*, **39**, 325.
6. Gamborg, O. L., Miller, R. A., and Ojma, K. (1968). *Exp. Cell Res.*, **50**, 151.
7. Dische, Z. (1962). In *Methods in carbohydrate chemistry* (ed. R. L. Whistler and M. L. Wolfrom), pp. 488–94. Academic Press, New York.
8. Evans, D. A. (1989). *TIG*, **5**, 46.
9. Negutiu, I. (1990). In *Plant cell line selection* (ed. P. J. Dix), pp. 19–38. VCH, Weinheim.
10. Atanassov, A. and Brown, D. (1984). *Plant Cell Tiss. Organ Cult.*, **3**, 149.
11. Ireland, K. F. and Leath, K. T. (1987). *Plant Dis.*, **71**, 900.

5

Plant regeneration via embryogenic suspension cultures

JOHN J. FINER

1. Introduction

There was a time when plant regeneration from most of the agronomically important crop species was thought to be impossible. Cereals and small grains, as well as important dicotyledonous species such as soybean and cotton, appeared quite unresponsive to tissue culture manipulations. Some species such as tobacco (*Nicotiana tabacum*) and wild carrot (*Daucus carota* 'Queen Anne's Lace'), on the other hand, could be placed *in vitro*, manipulated in a number of ways, and then regenerated to form whole plants with a simple change in the supply of growth regulators (1, 2). These model species had one thing in common, the ability to form whole plants from a totally undifferentiated state (callus). Any species that did not respond to tissue culture manipulations in a manner similar to the model plants was termed 'recalcitrant'. Undifferentiated tissue or callus could be produced from the 'recalcitrant' species but regeneration from this tissue was virtually impossible. Although the model species have been and will continue to be valuable in plant cell biology research, it is unfortunate that such a large effort was placed in attempting to force many plants to respond to the same manipulations as did the model plants. In the end, the responses of the early model species *in vitro* may be the exceptions rather than the norm.

Due to the value of the agronomic species, there has been a strong effort to develop tissue culture systems for these crops. A new type of callus with a smooth, rounded surface and containing cells with a dense cytoplasm was observed. Although not completely undifferentiated, these meristem-like cells proliferated slowly and plants could be regenerated from this tissue. Indra Vasil (U. of Florida), Bob Conger (U. of Tennessee), and Ed Green (U. of Minnesota), along with their students and colleagues, were the main contributors to the early work elucidating the potential of this 'embryogenic callus' in grasses. Similarly, Jerry Ranch (Pioneer) and Norma Trolinder (Texas Tech) contributed much to the understanding of embryogenesis in the agronomic dicots soybean and cotton respectively. Based on theirs and others' work, progress then followed on the development and refinement of embryogenic suspension culture systems.

Although the literature is full of reports of embryogenic callus and suspension cultures, it can be difficult to describe accurately a callus' phenotype, or, alternatively, to determine the fate of a piece of tissue based on the readings or photographs in a descriptive study. The best approach to gaining an understanding of embryogenesis is experimentation and experience. Once some experience is gained with one embryogenic system, it should not be difficult to perform parallel studies with another plant system. In fact, most embryogenic cultures look similar and respond to the same stimuli. These similarities become very apparent after working with many different systems.

This chapter describes the common features of embryogenesis systems and focuses on embryogenic suspension cultures. Protocols are given for initiation and regeneration of embryogenic suspensions of agronomically important monocot and dicot species; these protocols may be used in transformation studies with these crops.

2. Production of embryogenic suspension cultures

Embryogenesis is the process of embryo initiation and development. For zygotic embryos, embryogenesis starts at zygote formation, ends at seed maturation, and marks the beginning of the sporophytic generation of the life cycle. During embryogenesis, shoot and root meristems are initiated, the morphological pattern of the plant is determined, and carbohydrates, lipids, and proteins accumulate (3). During somatic embryogenesis, an embryo (similar to the zygotic embryo) containing both shoot and root axes, is formed from somatic plant tissue. An intact plant rather than a rooted shoot results from the germination of these somatic embryos.

For ease of explanation, the process of embryogenesis can be divided into four different stages. These are initiation or induction, proliferation, development or maturation, and finally germination (*Figure 1*). During initiation, cells are induced to form somatic embryos; during proliferation, these induced cells undergo multiplication with very little or no maturation of the embryogenic tissue. Although not entirely appropriate, germination is included in the process of somatic embryogenesis here because the dormancy associated with zygotic embryogenesis is often absent from somatic embryogenesis. In addition, many who work in the area of somatic embryogenesis also study germination of the somatic embryos. The proliferation stage will receive the greatest emphasis in this chapter. It is the least studied stage but is the most critical for embryogenic suspension culture work.

2.1 Initiation of embryogenic suspension cultures

For all tissue culture regeneration systems, there are numerous factors that affect tissue culture responses. These include the effects of media addenda

induction proliferation development germination

+auxin   -auxin

Figure 1. Schematic of four different stages of monocot somatic embryogenesis.

such as growth regulators, nitrogen sources, carbon sources, vitamins, and inorganic and complex organic addenda. In addition, the genetics and physiology of the starting or 'explant' material can have a major effect on the tissue culture response.

2.1.1 Explant status

Aside from the effects of various medium addenda, of equal or greater importance is the status of the explant. This includes the tissue type, physiological state of the tissue, and the genetic make-up of the donor plant. In most cases, the best explant for studies on induction of somatic embryogenesis is the immature zygotic embryo. This tissue is already embryogenic in nature and apparently requires less nurturing than other somatic tissues to elicit an embryogenic response. The zygotic embryo was not initially recognized as the best explant material for embryogenesis studies due to the relative plasticity of wild carrot (*Daucus carota*), the model for somatic embryogenesis studies. Almost any explant is adequate to induce embryogenesis in wild carrot.

The use of the zygotic embryo and description of a 'response window' for that zygotic embryo was first described for maize by Green and Phillips (4). Not only was it critical to use the zygotic embryo as the explant, but that embryo would only respond maximally if it were between 0.5 and 2 mm in length. The window for the zygotic embryo response was also observed with many cereals (wheat, rye, oats) as well as with other plants (5, 6). With some plants, explant tissue other than the proper staged zygotic embryo can produce somatic embryos, but the zygotic embryo is still considered to be the best overall explant for induction of somatic embryogenesis.

Beyond the physiological status of the explant, the next question is why are explants more responsive from some species than others? Even within a species, some cultivars are highly responsive while others do not give an embryogenic response. Are there 'regeneration' genes that can be isolated and studied? Ray and Bingham (7) developed alfalfa lines that were highly responsive to induction of embryogenesis using conventional breeding. The genetics of the embryogenic response was complex and multigenic. Recently, Armstrong *et al.* (8), using RFLP analysis, identified regions of the maize chromosomes that apparently carried genes that were associated with a high embryogenic response. A similar approach has been used for mapping genes involved in androgenic embryogenesis in maize (9). Although some of the genes that are involved in the embryogenic response have been mapped in maize, these genes have not yet been cloned and studied.

2.1.2 Media addenda

For somatic embryo induction and proliferation, you must supply an auxin, (usually 2,4-D) to the tissue. The levels of auxin that are used for induction of embryogenesis are often fatal to the intact dicot plants used for somatic embryogenesis experiments. It is intriguing that a group of herbicides (2,4-D, picloram, dicamba) contains the most effective inducers of somatic embryogenesis. Although there has been some effort to understand the molecular biology of embryogenesis (10), and some auxin-inducible genes have been cloned and studied (11), the specific role of auxin in the somatic embryo induction process remains unknown. It is clear that auxin application results in production of ethylene, which in turn can inhibit induction of embryogenesis. Inhibitors of ethylene production (aminovinyl glycine, AVG) and ethylene action (AgNO$_3$) can be used in tissue culture to enhance the embryogenic response in maize (12, 13). Benefits of ethylene inhibitors have not been documented in dicot embryogenesis systems.

The importance of nitrogen in somatic embryogenesis is without question. Tissue culture media usually contain nitrate and ammonia; both are available to the plant tissue at very high concentrations. These compounds are usually not limiting for growth in culture. Effects of nitrogen on somatic embryogenesis are more often seen through the addition of certain amino acids (14). It is unclear if these amino acids are beneficial for embryo initiation, proliferation, or development. It is interesting that the levels of the amino acids that affect somatic embryogenesis (proline, glutamine, asparagine, alanine) are enhanced during times of plant stress and are a major component of seed and vegetative storage proteins.

2.2 Proliferation

Although an understanding of embryo proliferation is critical for manipulation of somatic embryogenesis, this phase has not been well studied. In the rush to claim recovery of plants from somatic embryos, the benefits of somatic embryogenesis are ignored. These benefits include generation of large amounts of uniform embryogenic tissue for development studies, and large scale plant recovery, as well as transformation research (protoplasts, electroporation, particle bombardment), and *in vitro* selection of mutants.

Proliferation of embryogenic tissue can be obtained and studied using either semi-solid or liquid media. Although one would assume that it is more difficult to work with proliferative liquid culture systems, this is not the case. The following are the benefits of using liquid media as opposed to solidified media:

- more rapid growth
- more efficient selection
- fast and more uniform response to media manipulations
- easy visualization of tissue quality

The first three benefits are based on the high tissue-to-medium contact in liquid medium. In a liquid medium, the embryogenic tissue is bathed in medium while with solid medium, an embryogenic callus 'sits' on top of the medium. With callus, all compounds used by the plant cells must diffuse through the medium and pass through the tissue that is in direct contact with the medium. Gradients are established and, if embryos are formed on the top of a callus, it is not known what concentration or ratio of growth regulators was responsible for that response in that tissue. If embryos or embryogenic tissues are buried in a callus, they cannot be observed easily. In a liquid medium, the callus will dissociate and the embryogenic tissue will be free-floating. Since embryos and embryogenic tissues are small, you need a good quality inverted microscope for proper observation and evaluation of embryogenic suspension cultures. Microscopes having condensers and light sources that can be raised and/or rotated so that there is sufficient space to place a flask on the microscope stage are preferred. Direct observation of embryogenic tissue in liquid culture is effective, rapid, and simple once the observer becomes familiar with the morphology of high quality embryogenic tissue.

2.2.1 Morphology of embryogenic suspension culture tissues

Embryogenic suspension culture tissue from most higher plants is very similar in morphology when observed using an inverted microscope. Embryogenic suspension cultures do not exist as cultures of proliferating single cells. Although there are a few cases where single, isolated cells from embryogenic suspension cultures can form embryos directly (15), embryogenic cultures for the most part proliferate as aggregates of embryogenic tissue. Single cells can slough from these aggregates but these cells are more likely non-embryogenic and probably have low survival rates.

The cells that make up the aggregates of embryogenic tissue are typically small (~20 μm diameter), isodiametric, and densely cytoplasmic with small vacuoles (*Figures 2, 3, and 4*). This type of description pervades the literature and refers not to free-living single cells but to the cells that make up the aggregates of embryogenic tissue. The embryogenic aggregates (not cells) are dense, yellow to brown in colour (when observed with an inverted micro-scope), and have a relatively smooth surface (*Figures 2, 3, and 4*). The surface characteristics of these clumps result from the isodiametric morphology of their cells. If the surface of a cell aggregate is rough and elongated cells (banana-shaped) protrude from the surface (*Figure 3A, black-filled arrows*), the aggregate is either non-embryogenic or contains non-embryogenic tissue.

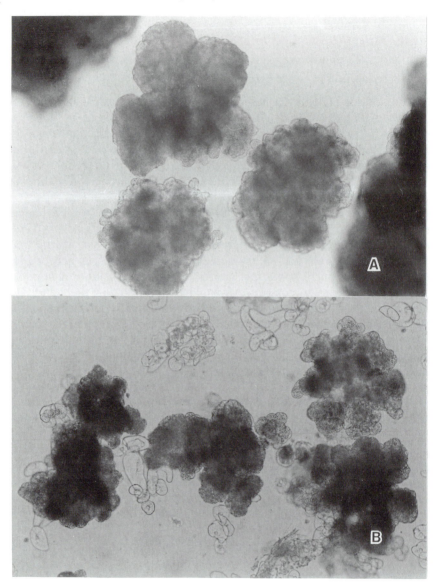

Figure 2. (A) Relatively pure, high quality embryogenic suspension culture tissue of cotton (*Gossypium hirsutum*). (B) Newly established embryogenic suspension culture of bentgrass (*Agrostis palustris*). Note the presence of both dense embryogenic clusters and non-cytoplasmic, translucent non-embryogenic aggregates.

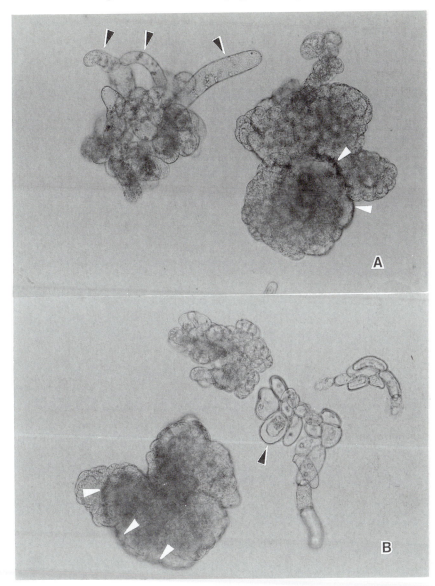

Figure 3. Embryogenic suspension culture tissue of maize (*Zea mays*). The *white arrows* show dark diffraction rings where the clusters are out-of-focus. (A) *Black-filled arrows* show non-embryogenic cells protruding from a cell aggregate. (B) *Black-filled arrow* shows a non-viable cell.

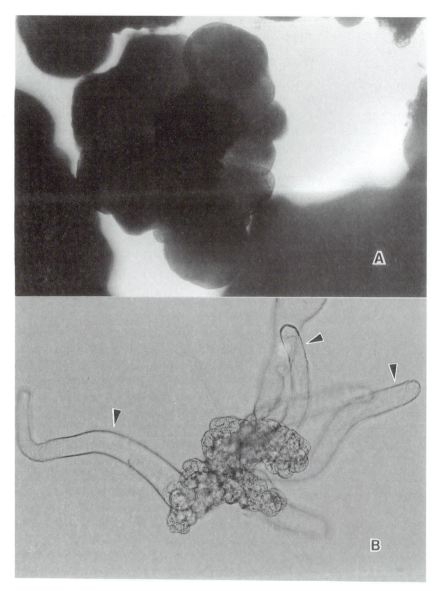

Figure 4. (A) Large cluster of dense, embryogenic soybean (*Glycine max*) suspension culture tissue. (B) Embryogenic suspension culture tissue of white pine (*Pinus strobus*). *Arrows* show suspensor cells attached to dense embryonal head.

Cells that contain a reduced or shrunken, irregular cytoplasm are probably not viable (*Figure 3B, black-filled arrow*). Isodiametric cells (embryogenic) will form a clump that has a smoother surface. The exception to the observation of a smooth surface for embryos is in the gymnosperms (*Figure 4B*).

Suspensor cells (*Figure 4B, arrows*), which are always associated with the embryo-proper in embryogenic gymnosperm suspensions, are elongate and somewhat vacuolated. Suspensor cells appear to be very similar to non-embryogenic cells but the embryonal head or embryo-proper is morphologically identical to the embryogenic clumps described above.

The surface of embryos at late developmental stages is very smooth from the presence of an organized epidermal layer. Proliferative cultures can contain globular and possibly later-staged embryos if the cultures are newly initiated or if the medium composition is not adequate to prevent embryo development. With soybean (16), fine proliferative cultures were never obtained and the cultures proliferate as relatively large (up to 8 mm diameter) clumps or masses of globular embryos (*Figure 4A*).

If developing or larger embryos can be observed in liquid medium, they should always be attached basally. This means that the smooth apical surface should protrude out from a central area. Cultures that contain clumps of tissue that have roots protruding outwards from a central core are probably root cultures. Since root formation in tissue culture is fairly common, one should critically examine the reports of 'embryo germination' with root formation but no shoot elongation.

The density of the embryogenic aggregates results from the cytoplasmic nature of the constituent cells. Non-embryogenic cells are more vacuolated and clusters of non-embryogenic cells are much less dense (*Figures 2B, 3A and B*). The density and colour of the aggregates reflect the growth rate or status of the tissue. Embryogenic aggregates that are rapidly proliferating are yellow to light brown while aggregates that have reduced growth appear very dark and are no longer translucent. If rapidly proliferating embryogenic aggregates are observed under the microscope and taken out-of-focus, they will exhibit a dark ring of diffraction and will appear almost 'oily' in appearance (*Figure 3, white arrows*). The colour and density of embryogenic aggregates can be used rapidly and precisely to gauge the quality of the embryogenic material in suspension culture. The effects of various media on the quality of embryogenic tissue can be accurately determined and liquid cultures that contain the highest amount of embryogenic material can be preferentially subcultured.

2.2.2 Preferential or selective subculture

Through preferential subculture, it should be possible to maintain embryogenic cultures for long periods of time (years) without a loss in embryo-forming capability. Plant regeneration capacity and fertility of regenerated plants usually decline in long-term embryogenic suspension cultures but the cultures do not turn non-embryogenic. Rather, the non-embryogenic tissue simply out-competes embryogenic tissue without selective subculture practices.

Selective or preferential subculture here has two meanings. The first refers to the selection of whole cultures that have the greatest proportion of proliferating, high quality embryogenic aggregates compared to non-embryogenic cells and debris. Flasks that contain the highest quality tissue are preferentially subcultured. With repeated subculture, this may narrow the genetic base of the suspension culture, but the quality of the suspension culture can be rapidly improved. The second type of selective subculture involves the isolation and purification of embryogenic tissue from liquid cultures. Embryogenic tissues can be identified with the aid of a microscope and subcultured into fresh medium. You can be extremely selective using this method by picking only a few pieces of the highest quality tissue for sub-culture. This type of selective subculture is not too different from isolation and maintenance of embryogenic callus cultures but you can subculture smaller pieces of tissue and this tissue can be more critically evaluated for subculture using the characteristics outlined above.

The 'low inoculum rule' for initiating and maintaining embryogenic suspension cultures is based on the ability of embryogenic tissue to survive and proliferate at very low inoculum densities in liquid culture. The main benefit of low inoculum culture is that very small amounts of tissue are required to initiate and maintain an embryogenic suspension culture. If you need to evaluate many different liquid media and high quality embryogenic callus tissue is limiting, you can use extremely small amounts of tissue (from 10 to 50 mg) to initiate liquid cultures. In addition, a very small amount (one or two aggregates) of proembryonic tissue from a mixed embryogenic and non-embryogenic suspension culture can be used for subculture.

A second benefit of low inoculum subculture of embryogenic tissue is that the non-embryogenic tissue does not survive subculture whereas the embryogenic tissue continues to proliferate regardless of cell density. In a pure embryogenic culture, non-embryogenic tissue can come from embryogenic tissue as sloughed cells and cell clusters. At high inoculum, this non-embryogenic tissue may interfere with embryogenic cell growth or proliferate but, at low density, these cells and cell aggregates have little effect and proliferate slowly if at all.

A third advantage of low inoculum culture is that the subculture period is extended. With rapidly growing cells in culture, the doubling time may be around three days, therefore cultures can be divided or split every three days if desired. A three day subculture regime will provide large amounts of rapidly proliferating tissue, but this is labour intensive and is quite inconvenient. If you use less tissue for initiation and maintenance of embryogenic suspension cultures, the subculture period can be lengthened to every one, two, or even four weeks. Subculture of approximately 100 mg (approx. 100 µl packed cell volume) of tissue every one to two weeks is both convenient and beneficial, providing adequate amounts of rapidly growing, high quality, embryogenic suspension culture materials.

The low inoculum rule can be easily verified. Take a small amount of high quality embryogenic tissue for initiation or maintenance of suspension culture. If the medium is not adequate for growth of embryogenic tissue (low or high inoculum) or if the tissue is poor quality (non-embryogenic or large embryos), you will not see a low inoculum effect. Perhaps the best place to start experimenting with low inoculum cultures is with newly established cultures and a medium that is known to support embryogenic growth. A series of subcultures comparing low and high inoculum will be a small effort and can save much time and effort in the long-term.

The quality of the embryogenic callus used for initiating suspension cultures cannot be overemphasized. Simply put, the higher the quality of the callus, the faster the suspension will become established. Without the proper starting material, it can be difficult to impossible to establish a high quality embryogenic suspension culture.

2.3 Methods for establishing an embryogenic suspension culture

Establishment of an embryogenic suspension culture may take from one to 12 months. Any callus which contains embryogenic sectors can be utilized but tissue that contains rapidly growing proembryonic material is best. The general methods used to initiate and maintain embryogenic suspension cultures are similar across genera, and there are some guide-lines that can be followed regardless of plant type. As with most procedures, there are exceptions. Below, a general procedure for initiation and maintenance of embryogenic suspension cultures is presented. This is followed by a description of five different systems that should provide information on most of the different situations that may be encountered with induction of liquid cultures.

In most cases, tissue that is already embryogenic (embryogenesis has been induced) is used to initiate liquid cultures (specific details on the induction of embryogenic callus are given in *Protocols 1–5*). You can use either small (100 mg) or larger amounts (1–2 g) of tissue to initiate suspension cultures. However, when using large amounts of tissue, you must closely watch cultures to prevent overgrowth of non-embryogenic or embryogenic tissue. At high density, either non-embryogenic tissue out-competes or outgrows embryogenic tissue or embryogenic tissue will start to senesce, a process that is sometimes difficult to reverse. You must also be cautious in distinguishing between embryo proliferation and development. A new culture may appear to be proliferating but the new tissue may be embryos that arise from pre-existing proembryos. Once embryo development is triggered, it is also a difficult process to reverse. Embryo development is distinguished by the presence of a smooth surface on the developing embryos when viewed using an inverted microscope.

For induction and maintenance of liquid cultures, 125 ml baffled deLong flasks are preferred. These flasks are relatively expensive but well worth the investment. The baffles aid in agitation of cultures resulting in superior tissue dissociation and good culture aeration. It is possible to get sufficient mixing in non-baffled flasks if the agitation speed is increased and/or the medium volume is reduced. Place 33–50 ml of medium in a 125 ml deLong flask and agitate at 130–150 r.p.m. A 125 ml Erlenmeyer flask agitating at 150 r.p.m. using 25 ml of medium may also be satisfactory. If, during culture with agitation, the cell clumps are not evenly distributed in the flask but remain in the centre of the flask, the agitation may be insufficient.

In addition to improved aeration, the deLong flasks are much easier to work with aseptically. The necks of deLong flasks are straight as opposed to those of Erlenmeyer flasks, which have slightly fluted neck openings. The advantage of the straight necks is that rigid plastic or metal caps can be used to close the flasks. These caps allow one-handed, one-step manipulations. To prevent any potential contamination problems during culture, seal the rigid caps to the flask using laboratory film. The film can be placed in such a position as to allow some gas exchange while keeping the cap firmly in place and sealing all large gaps.

2.3.1 Newly established suspension cultures

When embryogenic tissue is initially placed into liquid medium, non-embryogenic cells will be present. These cells are either present in the starting material or are sloughed from embryogenic material. Some cell cultures will also produce polysaccharides, which, together with the non-embryogenic cells, give the initiated suspension culture a cloudy appearance. It is important to make some initial observations (directly following initiation of liquid cultures) and follow the embryogenic cultures through their 'maturation'. The embryos and embryogenic tissue, if present, will first appear yellow to light brown under the inverted microscope but may become very dense or darken within a few days. Once the tissue has become adjusted to the liquid medium, the embryogenic tissue will start to proliferate and the new growth will appear yellow to light brown with the other visual characteristics described earlier. Some embryogenic tissue starts growth immediately following initiation of suspension cultures while in other cases, it can take one to four weeks for embryogenic tissue to resume growth in liquid culture. If the tissue remains dark and very dense for an extended period (when observed using an inverted microscope), this is an indication that the tissue is not rapidly proliferating. You can rapidly evaluate suspension culture media and response of embryogenic tissue to limited media by comparing clump density during the first few weeks following initiation of the suspension culture.

2.3.2 Purification of the suspension culture

If the media/tissue combination supports growth of embryogenic tissue, the 'young' embryogenic suspension culture will still contain a mixture of different cell types and debris. Before embryogenic cultures can be used for physiological studies or transformation work, the culture needs to be purified of the non-embryogenic tissue. There are at least four basic means of 'cleaning' a suspension culture of non-embryogenic cells and debris. Cleaning can be a continual process with some systems and these procedures must be used throughout the maintenance or subculture period. The choice of methods is dependent on the nature of the suspension cultures and suspension culture tissue. If the clumps of tissue are large, it certainly would not be appropriate to use a narrow-mouth pipette to transfer the tissue. If there is a lot of debris in the suspension, it may be difficult to separate clusters by filtration. Three general methods which can be used to purify embryogenic suspension cultures are described below.

i. Density transfer
This method uses density to separate physically the lighter non-embryogenic cell clusters, cells, and debris from the denser embryogenic clumps. Pipette 1–5 ml of a young embryogenic culture using a 10 or 25 ml wide-mouth glass or plastic pipette. Gently agitate the flask either just prior to (5–10 sec) or during the pipetting to resuspend the cell clusters. Then remove the pipette from the flask and place the tip in fresh medium in a new flask (without actively discharging the pipette). The denser embryogenic clusters will fall out of the pipette while the non-embryogenic cells and debris will remain in the pipette. Sometimes, if there is a lot of embryogenic tissue or the operator is not rapid enough, the embryogenic tissue will collect in the tip and temporarily clog the pipette. If this happens, take up a small amount of fresh medium (0.5 ml) into the pipette to resuspend the 'pellet'. Alternatively, gently tap the pipette to loosen the embryogenic tissue. This simple technique is very efficient for the separation of embryogenic and non-embryogenic tissue. You can selectively subculture small clumps by allowing the larger clumps to settle to the bottom of the flask prior to removal of embryogenic tissue using a pipette. This procedure, as with most clean-up protocols, does not result in instant success. From four to seven repetitive transfers are required before the culture is purified. Carefully watch the subculture period so that the cultures are not permitted to become dense with cell clusters and debris. As stated before, embryogenic suspension cultures that become overgrown will start to senesce and produce non-embryogenic tissue.

ii. Repetitive frequent transfer
In this method, frequently remove and transfer aliquots of the young suspension, usually at weekly intervals. At low density subculture, embryogenic tissue can survive this rapid subculture whereas non-embryogenic tissue does not survive. With frequent transfer, cellular debris and non-embryogenic tissue are simply diluted with each subculture.

iii. Manual selective transfer
You can physically select the aggregates of choice using either fine forceps and a good dissecting microscope or a pipette and an inverted microscope;

both methods must be performed in a sterile environment. This technique requires proper identification of embryogenic material and a steady hand to transfer small pieces of tissue in liquid without damaging that tissue. When using forceps to transfer a small piece of tissue, use one hand to hold the forceps as you normally would while the other hand can be used to stabilize the instrument. To stabilize the forceps properly, the index finger of the stabilizing hand should be touching the other index finger, thumb, and two prongs of the forceps at the same time. It should be possible to maintain a 0.5 mm gap between the forceps prongs using this technique. You can use a 1 ml pipette to transfer small pieces of tissue from a Petri dish using an inverted microscope. The small opening of the pipette is very large when viewed under an inverted microscope and, with some practice, it is possible to pick out single clumps of tissue less than 100 μm in diameter. It is difficult to obtain large amounts of proliferative embryogenic tissue using physical selection. Therefore, this method takes advantage of low inoculum subculture of embryogenic tissue.

2.4 Development and maturation of somatic embryos

For embryo development, the proliferative embryogenic tissue is transferred to a medium containing either no or low auxin and sometimes ABA, high sugar, or amino acids (see *Protocols 1–5*). The effect of ABA on somatic embryogenesis has received much attention in recent years. Ammirato (26) was the first to report 'normalizing' effects of ABA on somatic embryogenesis in caraway. He noticed that ABA suppressed embryo development and allowed more normal cotyledon formation. Crouch (27) demonstrated an accumulation of seed storage protein in *Brassica napus* following exposure of somatic embryos to ABA. Without this exposure to ABA, the somatic embryos failed to accumulate seed storage protein or develop normally. High sugar (6–18% sugar) or amino acids may substitute for this ABA effect as interactions between osmotic stress, ABA production, and accumulation of certain amino acids (glutamine, proline) are well documented.

Germination of developed somatic embryos can occur sporadically, with some embryos never germinating to form roots and shoots. Embryos may not germinate if they have not reached physiological maturity. Although embryos can be morphologically similar, they may be quite different physiologically. In most cases, it is beneficial to allow for an extended embryo development period to assure physiological maturity. With soybean, the somatic embryos turn from green to yellow (similar to zygotic embryos) as they mature. Embryos that develop over one to two months will be more responsive to desiccation and other dormancy-breaking treatments.

2.5 Specific protocols for plant regeneration via somatic embryogenesis

Procotols 1–5 describe the establishment of embryogenic cell suspensions and subsequent regeneration in orchardgrass, rice, maize, soybean, and cotton.

Establishment of an embryogenic callus culture of alfalfa is described in Chapter 6C, *Protocol 1*, and regeneration of caucasian bluestem through embryogenic cultures in Chapter 1, *Protocol 7*.

Protocol 1. Production and regeneration of embryogenic cultures of orchardgrass[a] **(*Dactylis glomerata* L.) (17)**

A. *Establishment of embryogenic cultures*

1. Grow *Dactylis* clones selected for high regeneration capability in the greenhouse.
2. Cut out several tillers including the root mass. Rinse off residual soil.
3. Cut off root mass leaving the basal meristem intact. Discard the root mass.
4. Split the youngest (innermost) leaves, longer than 4 cm, longitudinally along the midrib.
5. Sterilize the leaf tissue in 50% ethanol for 2 min and then rinse two or three times with sterile distilled water.
6. Slice the leaves into five 3 mm segments starting at the base.
7. Plate the leaf segments on to Schenk and Hildebrandt (SH, ref. 19) basal medium containing 30 μM (6.6 mg/litre) dicamba and 0.25% (w/w) Gelrite.
8. Culture the leaf segments in the dark at 25 °C. Embryogenic callus should form within one and a half to two weeks.
9. After three or four weeks of callus growth, remove approximately 0.5 g and inoculate into 50 ml of liquid SH medium containing 45 μM (10 mg/litre) dicamba and 4 g/litre casein hydrolysate in a 125 ml baffled flask.
10. Grow cultures in the dark on a gyratory shaker at 130 r.p.m. for four weeks.
11. Transfer to fresh medium using the 'sedimentation method' (20).[b]
12. Repeat step **11** every three to four weeks using the best quality flasks as judged by the abundance and quality of the embryogenic material (see Section 2.2).

B. *Regeneration of plants*

1. Plate 1 ml of the embryogenic suspension on SH medium containing 30 μM (6.6 mg/litre) dicamba.
2. Place cultures in the dark for two to three weeks.
3. Isolate and culture single somatic embryos (identified as solid, very pale structures against the cream coloured callus) and plate on growth regulator-free SH medium.
4. Place somatic embryos in low intensity light at 25 °C.

Protocol 1. *Continued*

5. When plantlets are about 3 cm tall, transfer to culture tubes containing growth regulator-free SH medium. When plants are 8–10 cm tall, transfer to soil, acclimate, and move to the greenhouse.

[a] Orchardgrass is one of the most responsive of the Graminae in production of embryogenic callus. Establishment of suspension cultures is routine but requires some experience. Best results will be obtained when using clones selected for high regenerative capacity, such as those developed by Conger *et al.* at the University of Tennessee (18).
[b] The sedimentation method is similar to the density transfer method (Section 2.3.2) but the larger cell clumps are allowed to settle prior to removal of the pipette from the stock culture. The result is the transfer of smaller clumps of embryogenic tissue.

Protocol 2. Production and regeneration of embryogenic cultures of rice (*Oryza sativa* L.) (from Horn, personal communication)

A. *Establishment of embryogenic cultures*

1. Surface sterilize mature, dehusked rice seeds by treating with bleach (5.25% $NaHClO_4$) for 30 min under constant agitation.
2. Rinse seeds three times with sterile distilled water.
3. Repeat steps **1** and **2** two more times to ensure disinfestation (mature seeds are able to tolerate harsh disinfestation treatments compared to other plant materials).
4. Plate seeds embryo side up on N6 medium (21) containing 12 µM proline, 20 µM 2,4-D (4.4 mg/litre) (N6–20 medium), and 0.25% Gelrite.
5. After three days, transfer the embryos that are not contaminated to fresh N6–20 medium.
6. After eight days, dissect the scutellar callus dome from the endosperm, root, and shoot. If non-embryogenic (NE) callus is present on the callus dome, remove it and discard.
7. Transfer the embryogenic callus to solid N6 medium for two to four weeks.
8. For initiation of embryogenic suspension cultures, inoculate 0.5–1.0 g fresh weight of pure embryogenic callus into 50 ml of liquid N6 medium containing 12 mM proline, 20 µM (4.4 mg/litre) 2,4-D, and 0.34 mM glutamine.
9. Examine the suspension cultures for quality. If necessary, perform selective subculture on the suspension culture material.

B. *Regeneration of plants*

1. For embryo development, collect the suspension culture material and rinse with liquid growth regulator-free N6 medium. Plate tissue on solid

N6 medium containing 12 mM proline, 50 μM ABA, 6% sucrose, and 2–5 μM (0.4–1.1 mg/litre) 2,4-D, and place in the dark for two weeks.

2. Transfer the callus to N6 medium containing 12 mM proline and 6% sucrose and place back in darkness for two to four weeks. Examine the cultures periodically for the appearance of somatic embryos. If few or no somatic embryos appear after four weeks, repeat steps **8** and **9**.

3. Remove mature somatic embryos and plate on MS medium containing 100 mg/litre inositol, 5 mg/litre thiamine HCl, 2 g/litre casein hydrolysate, 2.3 μM kinetin, 8 g/litre agar for germination.

4. Transfer plantlets containing shoots and roots to soil after they reach a height of 10 cm. Grow initially in the shade with moderate misting.

Protocol 3. Production and regeneration of embryogenic cultures of maize[a] (***Zea mays*** L.)

A. *Establishment of embryogenic cultures*

1. Harvest ears from plants about 12–14 days after pollination or when the embryos are 1–1.5 mm long and remove husks. Wash the ear thoroughly to remove silks, insects, and other large surface contaminants.

2. Insert pipette or long forceps into the basal end of the cob and put the cob into a sterile 500 ml beaker. Tall beakers usually work best. The pipette or forceps will serve as a cob handle for the rest of the procedure.

3. Pour 50% bleach into the beaker to cover the cob and soak for 30 min. Make sure that the cob does not float out of the bleach.

4. Pour off the bleach into a waste container and rinse the cob three times with sterile water to remove the bleach.

5. Cut off the top half of the kernels using a sterile scalpel. Kernels are not removed from the cob and the tops of 20–30 kernels are removed at a time. Change blades after every cob as they dull easily. The zygotic embryos are at the base of the kernel on the side wall toward the tip of the cob.

6. With a spatula, scoop out and discard the endosperm. Gently push on the side of the kernel wall to push the embryo up and out of the kernel.

7. Place the embryos, axes up (flat side down), on callus induction media.[b]

8. Allow embryogenic callus to form in the dark or low light at 25 °C for one to two weeks. Subculture high quality callus, as judged by friability and colour, on the same callus induction medium for two to four weeks with weekly transfers. These transfers require examination of the callus using a dissecting microscope and selective subculture.

9. Initiate embryogenic suspension cultures by inoculating 100–200 mg fresh weight of high quality embryogenic callus into 35 ml of liquid proliferation medium.[c]

Protocol 3. *Continued*

10. Subculture selected suspension cultures containing the highest proportion of high quality embryogenic material (*Figure 3*) using the density method (see Section 2.3.2). Subculture at low density is quite beneficial for these newly established cultures.

B. *Plant regeneration*

1. For somatic embryo development, plate the suspension culture on solidified induction medium for two to four weeks until an embryogenic callus is again formed. Then transfer the embryogenic callus to a growth regulator-free medium containing 6% sucrose with or without 12 mM proline. Somatic embryo formation and maturation can require from two to six weeks depending on the genotype and age of the suspension culture.

2. For germination of the mature somatic embryos, transfer tissue to a growth regulator-free medium with 2% sucrose.

3. Transfer plantlets to soil after they reach a height of 8–10 cm. Gradually expose plantlets to lower humidity in a growth chamber before moving to a greenhouse.

[a] The most responsive maize tissue for embryogenic callus and suspension culture work is obtained from the hybrid immature embryo from an A188 × B73 cross. Embryogenic callus and suspension cultures from this hybrid callus can be maintained on a simple medium containing MS salts, B5 (22) vitamins, 1–1.5 mg/litre 2,4-D, and 2% sucrose. Other lines have different media requirements such as inclusion of proline or casein hydrolysate, preference for other auxins such as dicamba, or increased sucrose concentration.
[b] The optimum callus induction medium depends on the genotype (23–25). Most of these media are based on N6 or MS salts and vitamins, 2% or more sucrose, and an auxin such as 2,4-D or dicamba.
[c] The liquid medium can be the callus induction medium without the gelling agent or may be substantially modified, again depending on the genotype.

Protocol 4. Production and regeneration of embryogenic cultures of soybean (*Glycine max* L. Merrill)

Soybean has proven to be the most unusual embryogenesis system of those described here. Initiation of proliferative embryogenic tissue requires use of extremely high 2,4-D levels and the embryogenic suspension culture is very clumpy.

A. *Establishment of embryogenic cultures*

1. Harvest pods from greenhouse grown soybean plants approximately 14 days post-pollination. The immature seeds should be about 4 mm in length and can be observed through the pod by back-lighting.

2. Wash the pods with warm soapy water and surface disinfect with 20% bleach for 20 min. Wash the pods five times with sterile distilled water.

3. Cut 4 mm off the end of the pod and peel back the pod to expose the immature seeds.[a]

4. Cut and discard the embryo axis by slicing 1–1.5 mm off the pointed end of the immature seed.

5. Put gentle pressure on the remaining portion of the seed to remove the two cotyledons. Place the cotyledons, flat surface-down, on the callus induction medium containing MS salts, B5 vitamins (22), 40 mg/litre 2,4-D, 6% sucrose, and 0.8% Nobel agar.

6. After six to eight weeks, remove the proliferative embryogenic material for initiation of suspension cultures.[b]

7. Place one small piece of tissue ($\leqslant$ 1 mm in diameter) in 35 ml of suspension culture medium containing modified MS salts (10 mM NH_4, 40 mM NO_3, ref. 16), B5 vitamins, 6% sucrose, and 5 mg/litre 2,4-D. Make an initial observation and observe weekly thereafter.

8. Monthly, remove only the high quality embryogenic tissue with forceps for subculture.[c]

9. For plant recovery, plate the clumps of embryogenic tissue on development medium containing MS salts, B5 vitamins, 6–12% maltose, and 0.2% Gelrite at 23 °C. After one month, isolate single embryos and plate on the development medium for one additional month.

B. *Plant regeneration*

1. After the embryos start turning yellow, place nine embryos in a dry 100 mm Petri dish for desiccation treatment.

2. Place the dish at 25 °C for two to three days.[d]

3. Transfer the embryos to a medium containing MS salts, B5 vitamins, 3% sucrose, and 0.2% Gelrite at 27 °C for germination.

4. After germination (one to two weeks), transfer the plantlets to a larger container such as a Magenta GA7 for further growth.

5. After the plant reaches the top of the container, transfer to a soil : sand : peat (1:1:1) mix and acclimate for two weeks before gradual exposure to ambient humidity and transfer to the greenhouse.[e]

[a] An immature soybean seed has a rounded and a more pointed end. The embryo axis is located at the pointed end.

[b] The tissue becomes proliferative when secondary somatic embryos grow from the apical surface of the primary somatic embryos. In cases where proliferation is vigorous, a whorl or rosette of secondary embryos is formed.

[c] The clumps of tissue will remain large (average size of 4 mm, *Figure 4A*) and a fine suspension culture will never be formed. Tissue that is spherical and consists of bright green, compact lobes is best.

[d] The embryos should wilt but should not lose too much water; smaller embryos will desiccate faster than larger embryos and severe desiccation is not beneficial.

[e] Unlike other germinating somatic embryos, soybean somatic embryos from suspension culture do not survive rapid transfer to a soil mix directly following germination. Harden them off *in vitro* as they increase in size prior to transfer to soil.

Protocol 5. Production and regeneration of embryogenic cultures of cotton (*Gossypium hirsutum*)[a]

A. *Establishment of embryogenic cultures*

1. Sterilize seed material; the sterilization technique is dependent on the status of the starting seed material.

   (a) If seeds are delinted, sterilize the seed with a 1–2 min dip in 70% ethanol, followed by treatment with 20% bleach for 20 min.

   (b) If seeds are delinted and coated with fungicide, rinse them four times with 95% ethanol prior to immersion in 20% bleach solution.

   (c) For seeds that are not delinted, place the seeds in concentrated H_2SO_4 for approximately 5 min or until the residual fibres are visibly removed and the seed turns black. Rinse the seeds carefully ten times with sterile water before bleach disinfection.

2. Rinse the seeds four times and place on growth regulator-free MS medium containing 3% sucrose and 0.8% agar. Germinate the seeds for one week at 27–31 °C with a light intensity of 30 $\mu E \cdot m^{-2} \cdot s^{-1}$.

3. Excise cotyledon sections and place pieces (approximately 3 × 3 mm) on medium containing MS salts, B5 vitamins (22), 3% glucose, 2 mg/litre NAA (or 0.5 mg/litre 2,4-D), 1 mg/litre kinetin, and 0.8% agar. Transfer the cotyledon pieces to fresh induction medium every week to prevent accumulation of phenolics in the medium.

4. After one month, transfer callus tissue to a medium similar to the induction medium but containing 3% sucrose.

5. After one additional month, transfer one piece of callus (250 mg) to 35 ml of suspension culture medium containing MS salts, B5 vitamins, 3% sucrose, and either 0.5 mg/litre picloram or 0.1 mg/litre 2,4-D.[b]

6. Observe cultures using an inverted microscope for the presence of embryogenic cotton tissue (*Figure 2A*). Subculture those flasks that contain embryogenic tissue using the suspension culture medium.[c]

7. After one additional month of proliferation, transfer the tissue to a medium containing MS salts, B5 vitamins, 3% sucrose, and 5 mg/litre 2,4-D. In this medium, the embryogenic tissues proliferate rapidly and the cluster size is reduced.

B. *Plant regeneration*

1. For embryo development, transfer the suspension culture tissue to the seed germination medium but containing 15 mM glutamine.

2. After one to two months of development, transfer the embryos to a

medium containing modified MS salts (no NH_4, two times NO_3), 1% sucrose, and 0.2% Gelrite. Nitrate treatment can be used to break dormancy in some seeds.

3. Transfer the germinating embryos to a 1 : 1 : 1 mix of vermiculite : perlite : peat under high humidity.

4. Gradually expose the plantlets to ambient humidity and move to the greenhouse.

[a] This procedure was developed for Coker lines of cotton. Other lines may respond to the manipulations but use of Coker 310, 312, or 315 is preferred.
[b] The callus used for initiation of cotton suspension cultures is the only tissue in this chapter that does not contain large, recognizable embryos. Callus should be yellow to cream coloured and friable.
[c] Over the first month of culture in liquid medium, it is not unusual for the culture medium to darken considerably.

3. Uses of embryogenic suspension cultures

Embryogenic suspension cultures have been utilized for many purposes. Their mitotic indices are quite high compared to callus cultures and most plant parts. This rapid division is important in the scale-up for cloning of elite germplasm. The suspension cultures are a good source for physiological studies and artificial seed production (see Chapter 6C) because the population of cells is more uniform than in callus, and somatic embryo development can be partially synchronized. The small size of the clumps makes penetration of growth regulators, inhibitors, and selective agents faster and more uniform, and makes scientific results less variable since gradients are less likely to complicate physiological studies. In many species of the Gramineae, embryogenic suspension cultures are the only source of protoplasts capable of cell division and subsequent plant regeneration. The relatively thin wall and high cytoplasmic density makes embryogenic suspension tissue ideal for cryopreservation work (see Chapter 7).

3.1 Desiccation and artificial seeds (see Chapter 6C)

Desiccation of somatic embryos serves two related purposes. First, in some species, desiccation breaks the dormancy associated with the somatic embryos. Secondly, desiccation promotes the accumulation of nutrients, such as storage protein, which will benefit the embryo when germination is allowed to proceed. Grape (29) and soybean (30) are two examples where the beneficial effects of desiccation on germination frequency have been shown.

The production of artificial seed (28) requires a large supply of normal mature somatic embryos. Embryogenic suspension cultures can provide the proper starting material for this type of work. As a mass expansion step, embryogenic suspension cultures are clearly superior to other tissue types with regards to growth rate. Although it is possible in some species (e.g. carrot) to produce artificial seeds from embryos developing while still in the liquid suspension phase, maturation for most somatic embryos requires plating on to a solid embryo development/maturation medium and then, at some later time, hand picking and processing (encapsulating) the somatic embryos.

3.2 Transformation

3.2.1 Protoplasts

Undoubtedly the most visible use of embryogenic suspension cultures has been for transformation in the economically important group of plants known as the Gramineae. I. K. Vasil and colleagues at the University of Florida showed that in the Gramineae, embryogenic suspension cultures were unique in their ability to release protoplasts capable of dividing at a reasonable frequency. Initially, this work met with much scepticism but it soon became apparent that embryogenic suspension cultures from a broad array of species such as pearl millet (31), Napier Grass (32), guinea grass (33), sugarcane (34), maize (35), orchardgrass (20), rice (36), and wheat (37) were capable of producing totipotent protoplasts.

These protoplast systems led to the first reports of recovery of transgenic Gramineae species such as maize (38), orchardgrass (39), and rice (40). Protoplast transformation is accomplished by direct gene transfer using one of two methods (see Chapter 3, Section 2):

- electroporation (41)
- polyethylene glycol (42)

Each of these protoplast transformation methodologies opens pores in the protoplasts allowing DNA to enter the cytoplasm. The difficulty with protoplast transformation systems is that protoplast isolation and culture is required (see Chapter 2). This is a laborious process and plant recovery from protoplasts may be extremely difficult in some cases.

3.2.2 Particle bombardment

The last several years have seen reports of alternatives to protoplasts for genetic transformation of monocot species. Recovery of transgenic plants via particle bombardment of embryogenic cells (43), apical meristems (44), and whole zygotic embryos (45, 46) is routine in certain laboratories. Whole tissue electroporation has also been reported as a viable alternative to particle bombardment (47). For transformation via particle bombardment and whole tissue electroporation to be successful, the target tissue must be competent to form germ-line tissues. It appears that cells that are actively dividing are more receptive to DNA introduction. Because apical meristems and whole embryos have been successfully used for some of these studies, it has been falsely assumed that the meristems simply elongate and the embryos germinate directly to form transgenic plants. This is far from the truth. To enhance recovery of transgenic tissue/sectors, the meristems and whole embryos proliferate, and any potentially transformed tissue can be either selected and/or multiplied. This proliferation takes the form of a shoot multiplication system for meristem transformation and embryogenic callus formation for whole embryo transformation.

Virtually all transformation methodologies show higher transformation efficiencies with suspension cultured cells than with callus, as judged by transient β-glucuronidase positive (GUS+) events. This holds for transfection in rice (48), use of SiC fibres in maize (49), and particle bombardment in maize (43), soybean (50), and cotton (51). Franks and Birch (52) bombarded regenerable suspension cultures, non-regenerable suspension cultures, and callus cultures of sugarcane and found that the regenerable suspension cultures showed over five and a half times the number of GUS+ events than was obtained from the callus tissue. The non-regenerable suspension culture tissue showed more than 41 times the number of GUS+ events than was obtained from the bombarded callus. The higher transient transformation rates are probably related to the higher growth rate of the suspension cultured cells compared to callus. Although embryogenic suspension cultures provide the most suitable starting tissue for most transformation work, use of this tissue has been hampered by a poor understanding of embryogenic suspension culture systems. Hopefully, this chapter will alleviate some of these concerns.

3.2.3 Fate of introduced DNAs

With all naked DNA transformation systems (e.g. direct DNA uptake with protoplasts, particle bombardment, whole tissue electroporation, SiC fibres), the fate of the introduced DNAs are similar. Once inside, the DNA somehow travels to the nucleus where it is incorporated into a site on one or more of the chromosomes. This DNA integration step is poorly controlled and a great deal of variability exists in terms of the level of gene expression and number of integrated gene copies. When plasmid DNA is introduced into plant cells, it recombines with both itself and plant DNA. Plasmid concatenation which is indicative of extrachromosomal homologous recombination has been reported with electroporated tobacco protoplasts (53) and bombarded cotton and soybean tissues (50, 51). Homologous recombination of plasmid DNA with plant DNA has been reported but the efficiency is very low (54). If the recombination process could be better controlled, copy number and therefore expression levels could be more accurately regulated. The variability in levels of gene expression in transgenic plant tissues has been somewhat reduced using scaffold attachment regions or SARs (55). These regions, which integrate with other transforming DNAs, may act by buffering the introduced DNA from repression by the native DNA. SARs appear to reduce variability and raise the level of gene expression between transformation events.

3.3 Cloning of elite germplasm

The inherent objective of cloning elite germplasm is to mass produce exact copies in as little time as possible. Somaclonal variation is not considered desirable in this type of endeavour. Since more clones can usually be produced with more time and tissue mass, it is useful to utilize a 'mass balance' calculation. This calculation details the steps of the system and provides the time needed as well as the mass expansion expected at each step. An example of such a calculation is shown in *Table 1* for rice. The primary stage of mass expansion is an embryogenic suspension culture initiated soon after callus induction and continued until the desired mass is achieved. Knowledge of the expected yield of somatic embryos per gram of tissue and the expected germination frequency of the somatic embryo is essential for making the decision as to the length of the suspension culture step. Mass expansion via the suspension culture step is clearly preferable to simple callus expansion since the growth rate is much faster in suspension cultures. Culture of tissue for relatively brief periods of time ($\leq$ 15 weeks) in suspension culture is not detrimental to subsequent embryo formation and plant germination. Beyond 15 weeks, some species have a tendency to produce fewer somatic embryos (see Section 2.3). Rice has been an excellent example of the potential that mass cloning brings to agriculture.

Table 1. Mass balance calculation for a typical rice cultivar (Horn *et al.* unpublished)

	Time elapsed	Mass
Embryogenic callus initiation	22 days	0.87 g [a]
Tissue proliferation	13 weeks	118.10 g
Embryo production	25 weeks	236 160 embryos
Plants	7 months	165 312 plants

[a] Per 20 seeds.

3.4 Physiological studies

Aside from carrot which appears to be a special case, there has not been a large number of basic physiological studies involving embryogenic cultures. Of particular interest would be studies comparing embryogenic and non-embryogenic cultures with regard to respiration, secondary metabolism, and phytohormone metabolism. There have been several such studies conducted with an emphasis on somatic embryo development and artificial seed. These have generally shown that ABA promotes embryo maturation with a concomitant increase in storage protein accumulation (56, 57).

Molecular analysis of embryogenic and non-embryogenic cultures have generated cDNA clones or protein profiles that are unique to embryogenic cultures. In many cases, the justification for this type of work has been to isolate a 'marker' for embryogenesis *in vitro*. Unfortunately, the physiological functions of these cDNAs and proteins have, in most cases, not yet been determined. In addition, the necessity for the isolation of markers (for identification of embryogenic versus non-embryogenic tissues) is questionable based on the information in this chapter. It is simply much easier to identify embryogenic tissue in suspension culture than to perform molecular analysis of that tissue to come to the same (or maybe a different) conclusion. The problem with a marker is that it may be difficult to determine at the molecular or biochemical level if a cluster of cells has become embryogenic. With visual selection however, it is possible to observe a rare event in a population of cells.

The value of studying gene expression during embryogenesis to understand and possibly control zygotic and somatic embryogenesis is without question. There has been some progress in understanding the molecular biology of embryogenesis using *Arabidopsis* embryogenesis mutants (58). This zygotic system should also provide some very valuable and exciting information relevant to the molecular biology of somatic embryogenesis.

3.5 Cryopreservation (see Chapter 7)

Cryopreservation is important for embryogenic suspension culture work for a number of reasons. In many cases, establishment of an embryogenic suspension culture can be a large effort, requiring input of much time and expertise.

(a) Once established, the suspension culture may be genetically stable for a short period of time.

(b) Embryogenic cells are highly cytoplasmic and hence contain less water for ice crystal formation.

(c) The small size of the embryogenic clumps allows quick and uniform penetration of the cyoprotectant solution. Penetration is not a problem with DMSO, which readily traverses membranes, but it is a factor for the osmoticum(s) such as sucrose or trehalose which are common constituents of the cryoprotectant solutions.

(b) The rapid cell divisons of embryogenic suspension culture material allows viable cells in a post-freeze environment to re-establish the culture in a short period of time.

Acknowledgements

The author wishes to gratefully acknowledge Dr Michael Horn, a good scientist and friend, for his numerous and unselfish contributions to this chapter. His input in this and other pursuits has been invaluable. Salaries and research support were provided by State and Federal funds appropriated to

6

Applied aspects of plant regeneration

6A. Micropropagation

STEFAAN P. O. WERBROUCK and
PIERRE C. DEBERGH

1. Introduction

Micropropagation is the true-to-type propagation of the selected genotype using *in vitro* techniques. In 1991 the world production of micropropagated plants was estimated to be 600 million plants, mainly in the categories of pot plants, cut flowers, fruit trees, and plants producing geophytes. The most frequently cloned crops are *Ficus*, *Syngonium*, potato, strawberry, *Spathiphyllum*, and *Gerbera* (1). The use of tissue culture for cloning plants is expensive. However, the introduction of automation and robotization to reduce manual labour is evolving quickly and will broaden the range of plants that can be propagated economically.

Micropropagation generally involves five stages, and this will be the guideline throughout this chapter:

(a) Stage 0: the preparative stage. The plant material for *in vitro* culture is prepared; the aim is to obtain hygienic and physiologically better adapted starting material.

(b) Stage 1: initiation of cultures. The aim is to obtain a reliable start. Multiplication is not important.

(c) Stage 2: shoot multiplication. The only function of this stage is to increase and maintain the stock; meristematic centres are induced and developed into buds and/or shoots.

(d) Stage 3: shoot elongation, root induction, and root development. Shoots are elongated (when they are too small to be manipulated) and rooted, or root induction takes place followed by root development *ex vitro*.

(e) Stage 4: transfer to greenhouse conditions. Most species require acclimatization in order to ensure their survival *ex vitro*.

2. Stage 0: the preparative stage (2)

2.1 Hygienic conditions

Stage 0 allows standardization of the growing conditions to yield hygienic explants, which need less aggressive sterilization. As a consequence, the plants will react in a more standardized way with the results being more reproducible. Some advice is given below.

(a) Grow the stock plants in a greenhouse.

(b) Use trickle irrigation or an irrigated sand bed; avoid overhead irrigation.

(c) Keep the temperature high (25 °C) and the relative humidity low (70%).

(d) Use thermotherapy for virus eradication (see Chapter 6B).

The minimum duration of a stage 0 regime has to be determined for each species. Three months seems to be sufficient for *Cordyline* spp., *Dracaena* spp., *Ficus* spp., and some *Araceae*.

2.2 Physiological conditions

With plant growth regulators or other treatments, one can try to improve the physiological fitness of the stock plants for *in vitro* culture:

(a) Plant growth regulator (PGR) treatment: spray, inject in stem, or apply during vase life.

(b) Constant photoperiod to control flowering.

(c) Rejuvenation by appropriate pruning, cascade grafting, bottom heating, or PGR treatment.

3. Stage 1: initiation of culture

3.1 Choice of explant

For most micropropagation work, the explant of choice is an apical or axillary bud. For only a limited number of plants, other explants are used, e.g. leaf pieces (*Ficus lyrata, Anthurium* spp., *Saintpaulia ionantha, Gloxinia*) or flower heads (*Gerbera jamesonii, Freesia*) on which adventitious buds are induced. It is always more risky to base a micropropagation scheme on adventitious buds since the chances of somaclonal variation will increase. When the objective is to produce virus-free plants from an infected individual, it becomes obligatory to start with submillimetre shoot tips. When choosing an explant, one should consider that:

(a) Aerial plant parts are less contaminated than underground parts.

(b) Interior plant parts are less contaminated than exterior parts.

(c) The smaller the explant, the smaller the risk of contamination.

(d) Regeneration capability is usually inversely proportional to the age and the size of the explant and to the age of the explant source.

3.2 Sterilization procedures

Protocol 1 describes an extensive sterilization procedure which is formulated for an axillary propagation scheme. It can be adapted for a scheme based on adventitious buds.

Protocol 1. Extensive sterilization procedure

1. Harvest shoots with three to six axillary buds.
2. Cut off the leaves, but leave the petiole stumps to protect the buds against sterilization damage.
3. Wash the plant material with tap-water and put it in a vessel. From now on work in a laminar flow hood.
4. Rinse in 95% ethanol for a few seconds and pour away the alcohol.
5. Add 0.1–1% $HgCl_2$ plus detergent (two drops/100 ml); pour away after 3–5 min. To avoid environmental pollution, retain the $HgCl_2$ and recover the Hg^{2+} by precipitation with ammonia.[a]
6. Rinse with autoclaved water.[a]
7. Add 7–15% NaOCl plus detergent (two drops/100 ml), pour away after 10–30 min.
8. Rinse three times with autoclaved water.
9. Cut off the damaged base of the shoot.
10. Cut nodal explants or isolate meristems and inoculate each in a separate tube.

[a] Steps **5** and **6** can often be omitted

3.3 Initiation media

Murashige and Skoog (MS) medium (3) has proven to be satisfactory for many crop plants. However, for some plants, the level of salts in the MS medium can be either toxic or unnecessarily high (e.g. *Ericaceae*). For these and also for many woody plant species, the woody plant medium (4) or the medium of Lepoivre (5) can be an alternative. Examples of initiation media are presented in *Table 1*.

3.4 Environmental conditions

Inoculated cultures are incubated in a culture room. *Table 2* lists the parameters which must be controlled.

Table 1. Examples of initiation media

Salts	Murashige and Skoog (3)
	Woody plant medium (4)
	Lepoivre (5)
Vitamins	Thiamine-HCl, 0.4 mg/litre
Inositol	100 mg/litre
Sucrose	2% (w/v)
Gelling agent	Agar 0.6% (w/v)
	Gelrite 0.12% (w/v)
	No gelling agent:
	• stagnant (paper wick)
	• agitated or rotated, 1 r.p.m.
pH	5.8
Hormones	Cytokinins: 2iP, BAP, Kinetin, Zea, or TDZ. Evaluate between 0–10 mg/litre; the concentration depends whether axillary or adventitious bud formation is expected
	Auxins (0–1 mg/litre)
	• axillary buds: IAA
	• adventitious buds: NAA or IBA
	Gibberellic acid (0–1 mg/litre filter sterilized) can be required for meristem cultures

Table 2. Environmental parameters in culture rooms

Temperature	18–28 °C, most often around 23 °C, night temperature can be 1 or 2 °C lower
Light:	
• Photoperiod	0 h, e.g. the first days after the initiation of meristem cultures, 16 h in most other cases
• Light quantity	Should be expressed as photosynthetically active radiation, a usual set point is 30 $\mu E \cdot m^{-2} \cdot s^{-1}$
• Light spectrum	Most often cool white fluorescent tubes are used
Relative humidity (RH)	The RH in the culture room is usually not controlled. In case of problems with hyperhydricity (vitrification) it may be necessary to cool the surface of the shelf to control the RH in the container, especially in stage 3 cultures

4. Stage 2: shoot multiplication

4.1 Media (6) (see *Table 1*)

The concept that organ differentiation in plants is regulated by an interplay of auxins and cytokinins should work as a guide-line when developing a medium for a new plant type. A higher cytokinin-to-auxin ratio promotes shoot formation and a higher auxin-to-cytokinin ratio favours root differentiation. This should not imply that, for adventitious shoot formation or axillary

130

branching, both hormones must be included in the medium. The exogenous requirements for hormones depend on the endogenous levels in the plant. In a number of cases a cytokinin alone is enough for optimal shoot multiplication.

Benzylaminopurine (BAP) is probably the most useful and reliable cytokinin and should be tested first for any new system. In the case of negative results, isopentenyladenine (2iP) can be tried. Zeatin (Zea) is not preferred because of the high purchase price. Generally 1–2 mg/litre cytokinin is adequate for most systems. Higher levels tend to induce adventitious bud formation. Thidiazuron (TDZ) is a very promising new cytokinin, which is effective at low concentrations (0.05–1 mg/litre). Because IAA is the least stable auxin in the medium, synthetic auxins such as NAA and IBA are preferred. For shoot multiplication their concentrations range from 0.1–1 mg/litre. Avoid 2,4-D because of its strong tendency to induce callusing.

4.2 Adventitious versus axillary bud formation

In all cases where mutations are not acceptable, only axillary bud development should be aimed for. Axillary buds are usually present in the axil of each leaf, and every bud has the potential to develop into a shoot. By growing shoots in the presence of a suitable cytokinin, apical dominance can be overcome and axillary buds grow directly into shoots. After four to eight weeks the original explant is transformed into a mass of ramified shoots or a cluster of basal shoots. When these miniature shoots or clusters are excised and planted on fresh medium (in which the cytokinin level could be raised), the shoot multiplication cycle can be repeated.

Using axillary shoots is not necessarily safe. An excessive dosage of a cytokinin or the choice of an inappropriate cytokinin can result in epigenetic off-types which usually can only be evaluated in the final product. A typical example is bushiness in *Gerbera jamesonii*, which is often the consequence of the use of the cytokinin BAP.

Some plants (e.g. *Solanum tuberosum*, *Dendrathema morifolium*, *Dianthus caryophyllus*) are propagated by growing elongating shoots which are subcultured by taking nodal cuttings.

Buds arising from any place other than the leaf axil or the shoot apex are termed adventitious buds. Usually adventitious organogenesis enables substantially faster multiplication. For each species one has to determine whether the adventitious bud system produces true-to-type plants. For many representatives of the *Gesneriaceae* (*Saintpaulia*, *Streptocarpus*, *Episcea*, *Aechimenes*, *Sinningia*), *Begoniaceae* (*Begonia rex*, *B. hiemalis*), and *Liliaceae* (*Lilium*, *Hyacinthus*) this system has proven to be reliable.

Most commercial micropropagation systems are a mixture of these organogenic pathways. In some cases, the establishment of a system where multiplication is only realized by axillary branching is probably impossible because *de novo* shoot production can not be excluded.

4.3 Number of subcultures (2)

The susceptibility to culture components and, as a consequence, the reaction of an explant, can change with the time in culture and the number of subcultures. On the same medium, stage 2 cultures, originally yielding axillary shoots, can produce abundant adventitious shoots after a number of subcultures. An unlimited number of subcultures increases the occurrence of epigenetic variation and mutation, which sometimes can only be recognized in the production field or in the greenhouse. Generally, 10–12 subcultures is considered to be a maximum.

5. Stage 3: shoot elongation and root induction or development (7)

5.1 Elongation

The relatively high levels of cytokinins in stage 2 produce a high number of shoots, which are sometimes difficult to manipulate. For most plants, good elongation can be obtained by transfer to a medium devoid of cytokinin. However, from an economic viewpoint, rather than transplanting single shoots, it is advisable to transplant clusters of shoots. The medium should allow good elongation of all shoots and therefore the dominance of only one shoot in a cluster will be avoided. To ensure this, cytokinin concentration can be lowered or a weaker cytokinin (kinetin or 2iP instead of BAP) can be used. It is also possible to add charcoal (0.3%).

5.2 Root induction

Adventitious and axillary shoots, which develop in the presence of a cytokinin, generally lack roots. Root development can be performed *in vitro* or *ex vitro*. Objections to rooting *in vitro* are:

(a) It is often very difficult to develop well functioning roots *in vitro*. After transfer to *ex vitro* these roots can die and new roots start to develop. Usually a delay in growth is observed.

(b) It is impossible to combine the optimum conditions for root initiation and root elongation in one medium (e.g. exogenous auxins are normally required for root induction but not for root elongation; they may even inhibit the process).

(c) The roots formed *in vitro* are usually damaged during transplanting, which considerably enhances the risk of root and stem diseases (e.g. *Fusarium*, *Pythium*).

(d) Shoots can be treated as cuttings. Striking cuttings is less labour intensive than manipulating plants with roots.

In a further attempt to save manual labour, instead of transplanting the shoots to fresh medium, liquid media can be added to established cultures (double layer technique) (8). For most species, auxins such as NAA or IBA (0.1–1 mg/litre), are required to induce rooting. The salt concentration and the application time are important factors which determine the success of this method. Most often a lower salt concentration is required, e.g. Knop's medium (9). Activated charcoal, added to the liquid medium, eliminates the residual effects of cytokinins by absorption. Higher sugar concentrations (3–4%) improve the rooting and the quality of the plants. It is possible to achieve both shoot elongation and root induction with one (liquid) medium. Especially in this stage, bottom cooling can be required to prevent hyper-hydricity. *Table 3* lists the culture conditions and media for stages 1, 2, and 3 for variegated cultivars of *Cordyline fruticosa*.

6. Stage 4: transfer to greenhouse conditions

The physiological and anatomical characteristics of micropropagated plantlets

Table 3. Stages 1, 2, and 3 media and culture conditions for variegated cultivars of *Cordyline fruticosa* (10)

Stage Aim	1 Initiation	2 Multiplication	3a Elongation	3b Root induction
Duration (weeks)	5–8	6	6	1
Light[a] ($\mu \cdot m^{-2} \cdot s^{-1}$)	30	30	3 weeks at 30 and 3 weeks at 100	100
Container	Tube	Vessel	Vessel	Vessel
Medium volume (ml)	20	100	+20[b]	+20[b]
Medium composition:				
• Macroelements	MS	MS	MS	10% MS
• Microelements	N and N (10)	N and N	N and N	N and N
• NaFeEDTA (mg/litre)	33	33	33	33
• Thiamine–HCl (mg/litre)	0.4	0.4	0.4	0.4
• Inositol (mg/litre)	100	100	100	100
• Adenine sulfate (mg/litre)	80	80	80	80
• NaH_2PO_4 (mg/litre)	170	170	170	170
• Sucrose (%)	2	2	4	2
• Agar (%)	0.6	0.6	0	0
• NOA[c] (mg/litre)	0.1	0.1	0	0
• IBA (mg/litre)	0	0	0	0.5
• Kinetin (mg/litre)	1–10	1–10	0.25	0

[a] 16 h cool white light (fluorescent tubes).
[b] Double layer application.
[c] Naphthoxyacetic acid.

necessitate that they gradually acclimatize to the environment of the green-house or field. Several techniques can be applied:

(a) The process of acclimatization can start *in vitro*. Bottom cooling reduces the relative humidity in the head space of the container and this can initiate the weaning process.

(b) Uncap the culture vessels and put them in the greenhouse several days prior to removal of the plants from the culture medium. Contamination of the medium does not become problematic unless plantlets remain in the open vessels for more than one week.

(c) Wash the agar thoroughly from the plants (it serves as a substrate for the growth of disease-causing organisms).

  i. When the shoots are rooted *in vitro*, transplant them carefully to reduce the risks of wounding.

  ii. In cases where the roots are newly induced, treat the shoots as cuttings and strike them directly.

  iii. When the roots have not yet been induced, dipping of the plant base in the rooting powder or solution before striking can be sufficient.

Unfertilized peat which is not wet is a good substrate. Rockwool plugs and other inert material are also useful.

Maintaining a high relative humidity for the first few days is critical. Equipment for maintaining high humidity includes: polyethylene tent, humidifiers (e.g. sprinklers, fog unit, ultrasonic evaporator), special boxes. The finer the water droplets (10 μm or less), the better, as this avoids a too wet and consequently too anaerobic substrate. A screen in the greenhouse prevents temperature peaks and lowers the light intensity, which makes the transition easier.

Acknowledgements

We acknowledge the financial support of the I.W.O.N.G., Belgium.

References

1. Pierik, R. L. M. (1991). In *Micropropagation: technology and application* (ed. P. C. Debergh and R. H. Zimmerman), pp. 155–63. Kluwer Academic Publishers, Dordrecht.
2. Debergh, P. C. and Read, P. E. (1991). In *Micropropagation: technology and application* (ed. P. C. Debergh and R. H. Zimmerman), pp. 1–13. Kluwer Academic Publishers, Dordrecht.
3. Murashige, T. and Skoog, F. (1962). *Physiol. Plant.*, **15**, 473.
4. Lloyd, G. and McCown, B. (1980). *Proc. Int. Plant Prop. Soc.*, **30**, 421.

5. Quoirin, M. and Lepoivre, P. (1977). *Acta Hortic.*, **78**, 437.
6. Bhojwani, S. S. and Razdan, M. K. (1983). *Plant tissue culture, theory and practice. Developments in crop science (5)*, pp. 502. Elsevier, Amsterdam.
7. Debergh, P. C. and Maene, L. J. (1981). *Sci. Hortic.*, **14**, 335.
8. Maene, L. J. and Debergh, P. C. (1985). *Plant Cell Tiss. Organ Cult.*, **5**, 23.
9. Knop, W. (1965). *Landwirtsch. Vers. Stn.*, **7**, 93.
10. Debergh, P. C. and Maene, L. J. (1983). In *Handbook of plant cell culture* (ed. P. V. Ammirato, D. A. Evans, W. R. Sharp, and Y. P. S. Bajaj), Vol. 5, pp. 337–51. MacGraw-Hill Publishing Company, NY.

6B. Virus-free plants

JOHN L. SHERWOOD

1. Introduction

Virus-free plants of many species and/or cultivars have been produced by culture of meristematic tissue (1). However, this method should not be considered routine. Specifics of protocols to eliminate virus from different plants vary widely (2).

2. Overview of methods

Three major requirements for production of virus-free plants are:

(a) a method to detect the virus;

(b) assessment of the type and extent of therapy required to eliminate the virus and maintain tissue viability;

(c) conditions and media for plant culture and differentiation.

2.1 Facilities and equipment

For heat therapy, an incubator or growth chamber is needed. The sophistication of the chamber depends on the requirements outlined in Section 2.
  For tissue culture, the following equipment is needed:

- Laminar flow hood with gas (Bunsen burner) or alcohol lamp
- Binocular dissecting microscope (magnification 10–40×)
- Illuminator with fluorescent lamp or fibre optics to minimize heat to the illuminated area

- Forceps and/or haemostats; scalpels and/or razor blades
- 75% to 95% ethanol; sodium hypochlorite solution (commercial bleach); detergent (e.g. Tween 20); sterile water
- Prepared tissue and prepared medium

2.2 Heat therapy

Heat therapy or heat treatment has long been used to rid plant material of infectious agents. All viruses and all plants do not react similarly to heat treatment. A balance must be struck between conditions that permit plant growth, but maximize virus elimination. Heat is generally applied by growing plants near 36–37 °C for days to months, but it can also be applied to material in culture. New growth may be free of virus and can be cultured or grafted on to other plants. Temperature, length of treatment, and intensity of illumination are conditions that are generally varied. Both duration of elevated temperature and length of illumination in a 24 hour period may be important. Heat therapy may not be successful, but is a useful treatment of material prior to tissue culture.

2.3 Selection of tissue and meristem culture

In a strict botanical sense, the meristem consists only of the cells in the apical dome. From a practical sense, it is very difficult to dissect and successfully regenerate the meristem. Use well nourished actively growing shoots. Success of regeneration from meristematic areas of either apical or lateral buds varies. The growth stage of the plant, the growing conditions, and season may influence the success of regeneration. Use material from plants grown as cleanly as possible to reduce the likelihood of cultured tissue being contaminated. Although virus-free plants have been obtained from callus cultures, the somaclonal variation in callus makes meristem culture preferable.

Meristems on shoots covered by developing leaves and leaf primordia are considered aseptic. Thus, the purpose of treatment of the shoot prior to dissection is to reduce the possibility of contaminating meristematic tissue during dissection. Treatment of the shoot depends on the contamination of and the sensitivity of the tissue.

Protocol 1. Isolation and culture of meristems for virus elimination

1. Dip tissue in 75–95% ethanol or 0.1–0.5% sodium hypochlorite (both amended with a drop of detergent (e.g. Tween 20) to increase wetability) for a few seconds to minutes, followed by several rinses in sterile water. In some cases disinfection by both treatments in succession may be needed.

2. Perform dissection in the laminar flow hood with the aid of the binocular dissecting microscope that has been wiped with ethanol. Using fre-

---

quently sterilized instruments, carefully remove the outer leaves and leaf primordia to expose the meristematic area. The meristemic area is easily injured and quickly desiccates. Be sure instruments sterilized by flaming are cool and that illumination of the working area does not dry out the tissue.

3. Once the meristematic area is exposed, remove it and place it on the surface of the medium in a culture tube. More than the meristem is usually taken, and a tip of 0.5 ± 0.2 mm is commonly used.[a]

4. Refer to publications on culture of the species and/or cultivar of interest for developing the appropriate medium (mineral salts, pH, vitamins, organics, hormones) and the culture conditions (light quality and amount, temperature) for shoot elongation and root development (see Chapter 1).

[a] Generally, the larger the size of tissue removed, the greater the success of regeneration. However, the larger the size of tissue removed, the less the success of regenerating virus-free plants. Meristematic areas are not free of virus. Virus in the tissue is eliminated during the culture process, by an unknown mechanism.

Heat treatment and/or meristem culture alone may not lead to culture of virus-free plants (3). Many chemicals have been added to media to try to enhance production of virus-free plants. The most commonly used with success is 1-β-D-1-H-ribofuranosyl-1,2-4-triazole carboxamide (ribavirin, virazole) (4) which, after sterilization by filtration, can be added to cooling media before solidification. The reaction of a species to ribavirin and subsequent success in producing virus-free plants varies. Concentrations ranging from 0–100 mg/litre and length of time in culture should be tested. Generally, as the concentration of ribavirin and length of time in culture increases the effectiveness of virus elimination increases. However, concentrations above 20–50 mg/litre reduce the rate of plant growth and can be phytotoxic.

3. Detection of viruses in cultured tissue

Following establishment of plants in soil from culture, they are tested for virus (5). Because plants may be infected, but assays for virus are negative, some prefer referring to plants taken through this process as being virus-tested rather than virus-free.

3.1 Detection by infectivity

An indicator host that produces characteristic symptoms after infection by mechanical inoculation can be very useful, especially if the virus has not been identified.

3.2 Detection of virus or virus inclusions

Light or electron microscopy are useful for looking for either inclusion bodies produced as a result of virus infection or for virus particles, respectively.

3.3 Detection of virus components

Several types of serological assays, nucleic acid hybridization, and the poly-merase chain reaction (PCR) are used to detect virus components. Their utility depends on the availability of antiserum, cloned probe, or specific primers (See Chapter 3C). Although these techniques are very sensitive, virus infection can be missed. As with the other assays, plants should be assayed several times during growth.

4. Conclusions

Each parameter may vary with the species and/or cultivar. A virus-free plant is not virus resistant. Care must be taken so that plants produced do not become infected. Virus-free plants can serve as a source for propagation of other virus-free plants.

Acknowledgement

Journal article No. 6288. Oklahoma Agricultural Experiment Station, Oklahoma State University, Stillwater, OK.

References

1. Wang, P. J. and Hu, C. Y. (1980). In *Advances in biochemical engineering* (ed. A. Fiechter), Vol. 18, pp. 61–99. Springer-Verlag, Berlin
2. Stace-Smith, R. (1985). In *Comprehensive biotechnology; the principles, applications, and regulations of biotechnology in industry, agriculture, and medicine* (ed. M. Moo-Young), Vol. 4, pp. 169–79. Pergamon, New York.
3. Kartha, K. K. (1986). In *Plant tissue culture and its agricultural applications* (ed. L. Withers), pp. 219–38. Butterworth, London.
4. Long, R. D. and Cassels, A. C. (1986). ibid., pp. 238–48.
5. Matthews, R. E. F. (1991). *Plant virology*. Academic Press, NY.

6C. Artificial seeds

D. C. W. BROWN

1. Introduction

The concept of artificial seed (also commonly referred to as synthetic seed or encapsulated embryos) has evolved over two decades as an alternate and potentially more efficient method to conventional micropropagation. The first schemes of direct fluid drilling of somatic embryos developed into techniques of encapsulating embryos in hydrated coatings and, more recently, into the idea of drying *in vitro*-derived embryos. The next stage of technique under development is the development of an effective coating to provide protection during storage and rehydration and the improvement in bioreactor production of embryos (see ref. 1 for detailed reviews). The paradigm being pursued in our research is a dry somatic embryo with a synthetic endosperm which would contain additives such as protein or lipid reserves, fungicides and/or *Rhizobium*, all of which would be protected with a synthetic coating to control rehydration and protect against physical damage during handling. The three protocols described here are based on the concept of producing viable, dried *in vitro*-derived embryos on a small laboratory scale. As shown in *Figure 1*, the protocol for alfalfa (*Medicago sativa*) is a flexible multi-step procedure which relies on simple and readily available equipment for laboratory scale production of artificial seed. Although the embryos used in each protocol described here are quite different in origin, the basic technique is the

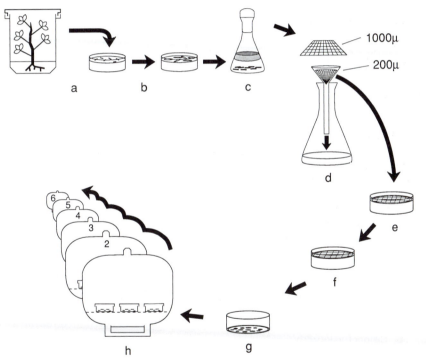

Figure 1. Protocol for laboratory scale production of alfalfa artificial seed. (a) Isolation of petiole explants from *in vitro* grown donor plants. (b) Callus proliferation and embryo induction on agar solidified medium. (c) Cell and embryo proliferation in suspension culture. (d) Embryogenic cell selection with nylon screens. (e) Embryo development on agar solidified medium. (f) Induction of desiccation tolerance with abscisic acid on agar solidified medium. (g) Embryo selection by size and/or colour. (h) Controlled desiccation by exposure of embryos to decreasing relative humidity conditions in sealed desiccators (as outlined in *Table 2*).

same for alfalfa petiole-derived embryos, *Brassica napus* microspore-derived haploid embryos, and *Triticum aestivum* immature zygotic embryos. Embryos are isolated or induced and grown *in vitro* to the proper stage of development, exposed to specific concentrations of abscisic acid for a particular period of time, and then dried to low moisture levels under controlled conditions.

2. Dried somatic embryos of *Medicago sativa*

Many explants can be used and will regenerate (2); however, *Protocol 1* uses 1 cm petiole explants. As the regenerability of alfalfa is genotype dependent, individual responsive plants must be selected (2). This particular protocol was optimized using a highly embryogenic clone A70–34 isolated from *Medicago sativa* cv. Rangelander.

Protocol 1. Production of dried somatic alfalfa embryos

1. Plunge whole petioles which have had two of the three leaflets removed into a saturated, pre-filtered solution of calcium hypochlorite for 10 min.[a] Agitate gently and periodically during the treatment. Wash three times with sterile distilled water.

2. Aseptically trim 1–2 mm from the end of each petiole, cut into approximate 1 cm sections, and float on sterile distilled water.

3. Transfer 1 cm petiole sections on to callus induction medium (B5h, *Table 1*). To avoid desiccation, float the petioles in a small amount of sterile distilled water in a Petri dish during cutting and while awaiting transfer.

4. Plate five to ten petioles on 20 ml solidified B5h medium contained in a 20 × 600 mm Petri dish wrapped with Parafilm prior to culture at 25 °C under a 12–16 h photoperiod of fluorescent light at 50 $\mu E \cdot m^{-2} \cdot s^{-1}$. Culture for 21–28 days until early globular-staged embryos are visible on the surface of the callus.

5. Aseptically transfer five to seven well developed calli (20–30 g) to 45 ml liquid B5m medium (*Table 1*) contained in a 125 ml Erlenmeyer flask (see *Figure 1*). Plug flasks with autoclavable foam plugs, cap with aluminium foil, and culture on an orbital shaker at 130 r.p.m. under the above conditions. Cell dispersion and growth should be evident within three days.

6. Select flasks with rapidly growing cell suspensions, 8–12 days old with a clear 'growth ring' of cells and young green globular embryos on the mid-upper portion of the flask.

7. Pass the cell suspension sequentially through a 1000 μm and 200 μm nylon screen, and retain the cells which accumulate on the 200 μm screen.

8. Transfer about 1 g of selected cell suspension to an aseptic 50 mm diameter 200 μm mesh nylon screen on 20 ml of solidified, BOi2Y

medium (*Table 1*) and spread evenly over the surface of the nylon screen (see *Figure 1*).[b]

9. In order to reduce secondary embryo formation and induce tolerance to desiccation, culture embryos on BOi2Y medium supplemented with 10^{-5} M abscisic acid by transferring the nylon screen containing the cell suspension and developing embryos on to the BOi2Y–ABA medium at day 11. The glutamine and sucrose incorporated in this medium are beneficial to embryo maturation.

10. After 7–14 days on ABA-containing medium and after embryos have acquired a cream-white colour,[c] harvest the embryos by collecting on a 200 μm nylon filter and washing off undeveloped cells and residual medium with sterile distilled water.

11. Transfer, without wrapping, Petri dishes containing selected embryos[d] spread on damp filter paper to a desiccator with a controlled relative humidity of 97% (see *Figure 1*).

12. Every 24–48 h, transfer the dishes to desiccators with progressively lower relative humidity levels,[e] as outlined in *Table 2*.

13. After desiccation, store dried embryos in the dark at room temperature and a relative humidity of 43%.

---

[a] The use of *in vitro* propagated plants, as shown in *Figure 1*, will reduce the risk of contamination and, in this case, the sterilization step can be eliminated. Alfalfa nodes can be propagated *in vitro* on basal MS medium (3) or on SH medium (*Table 1*) at 25 °C under a 12–16 h photoperiod. Light intensity is maintained at 50–125 $\mu E \cdot m^{-2} \cdot s^{-1}$.

[b] Embryos will develop on BOi2Y medium in a semi-synchronous fashion. The first globular embryos are visible three to five days after transfer to BOi2Y and large numbers of embryos are visible by day 21. A conversion frequency of about 70% can be achieved if 21–28 day-old cotyledon staged embryos are separated and cultured vertically on Murashige and Skoog medium lacking growth regulators. When cultured in this manner, the cultured embryos tend to form secondary embryos along the hypocotyl region of the embryo. This phenomenon is usually observed after day 18 on BOi2Y medium.

[c] As alfalfa torpedo and cotyledon staged embryos acquire desiccation tolerance they gradually loose chlorophyll and take on a cream-white colour.

[d] Embryos can be selected by developmental stages, size, or colour, or transferred as a population to a sterile filter paper (Whatman No.1), dampened with sterile distilled water, contained in a 60 × 20 mm plastic Petri dish. Care should be taken that embryos do not dry out quickly during manipulation in the laminar flow bench or that excess moisture is present on the filter paper. Petri dishes are covered but not sealed before transfer.

[e] Relative humidity levels are controlled by including specific saturated salt solutions at the base of each desiccator. Relative water loss from the plant material can be monitored with relative humidity gauges and samples transferred after 24 h or, when large samples are desiccated, when the relative humidity in the chamber approaches the theoretical value which represents atmosphere equilibration (see *Table 2*).

For rehydration, place dried embryos directly on to filter paper on MS medium (Chapter 1) lacking growth regulators or transfer the filter paper directly on to the medium. Germinating embryos can be assessed for survival one or two weeks after being placed on the medium by monitoring for shoot and root growth. Root germination only indicates ineffective or partial

Table 1. Media formulations for *Medicago* artificial seed protocol—additions are given as mg per litre final concentration

	SH$_f$	B5h	B5m	BOi2Y	BOi2Y-ABA
Base salts/vitamins	SH[a]	B5[b]	B5[b]	B[c]	B[c]
NH$_4$NO$_3$	1000				
CaCl$_2$·2H$_2$O		600			
Glutamine		800			7305
Glutathione		10			
Serine		100			
Adenine		1			
Inositol				100	100
Yeast extract				2000	2000
Sucrose	20 000	30 000	20 000	30 000	50 000
2,4-D		1.0	1.0		
Kinetin		0.2			
Naphthalenacetic acid			0.1		
Indole-3-acetic acid	0.02				
Isopentenyladenine	0.02				
Abscisic acid					13.2
Bacto agar (Difco)	10.0	9.0		10.0	10.0
pH	5.9	5.5	5.5	5.9	5.9

[a] Schenk and Hildebrandt medium (5), see Chapter 1.
[b] Gamborg *et al*. B5 medium (6), see Chapter 1.
[c] Blaydes medium (4).

Table 2. Relative humidity conditions used during desiccation

Desiccator	Saturated compound	Theoretical % RH (8–11)	Measured % RH
1	K$_2$SO$_4$	97.5	92
2	Na$_2$CO$_3$·10H$_2$O	87	92
3	NaCl	75.5	88
4	NH$_4$NO$_3$	62.5	79
5	Ca(NO$_3$)$_2$·4H$_2$O	50.5	52
6	K$_2$CO$_3$·2H$_2$O	43	48

induction of desiccation tolerance as our observations suggest that root meristems are more hardy than shoot meristems.

3. Dried microspore-derived embryos of *Brassica napus*

For obtaining donor plants, germinate seeds of *Brassica napus* in 4 inch peat pots and grow the plants in a growth chamber at 20/15 °C (day/night) with a

16 h photoperiod and 300 $\mu E \cdot m^{-2} \cdot s^{-1}$ light intensity. Five weeks after germination, when the plants begin to bolt, lower the growth chamber temperature to 10/5 °C (day/night). Lines and cultivars Topas, Westar, SCSN-1, Reston, Profit, and Jet Neuf have been reported to respond well using this protocol (7, 8).

Protocol 2.
Production of *Brassica* microspore-derived embryos

1. Surface sterilize flower buds (2.5–4.0 mm in length, showing minimal pedicel elongation and having a green to yellow-green colour) for 15 min in 7% calcium hypochlorite and wash three times for 5 min in sterile distilled water.

2. Transfer buds to B5 medium (5) modified to contain 13% sucrose (B5–13), contained in a sterile 50 ml beaker, and macerate with a glass pestle against the side of the beaker, to release the microspores.

3. Filter the homogenate through a 44 μm nylon mesh screen, transfer to a centrifuge tube (16 × 100 mm), cap, and centrifuge at 100 *g* for 3 min. Decant the supernatant.

4. Wash the microspores with fresh B5–13 medium and repeat the centrifuge step twice with the speed being progressively reduced to 75 *g* and 50 *g*.

5. Resuspend the microspores in filter sterilized NLN medium (7) at a density of 20 000–30 000/ml. Pipette a 1.2 ml sample of the microspore suspension on to 60 × 15 mm Falcon 1007 plastic Petri dishes containing 2 ml of NLN medium solidified with 0.35% agarose. Wrap the plates with Parafilm.

6. Transfer the plates to 32.5 °C, in the dark, for four days. Transfer to the 32.5 °C treatment as soon as possible as any substantial delay (2 h+) results in a reduction of embryo induction (9).

7. After four days of culture at 32.5 °C, transfer the dishes to 25 °C, in the dark, to allow embryo development.

8. To induce desiccation tolerance, between day 14 and 21, carefully pipette off the medium and add new filter sterilized NLN medium containing 10^{-4} M ABA to the culture dish. Add enough medium to bring the level up to that of the original level in each Petri dish. Culture for seven days in the light or dark at 25 °C.

9. At day 21, select embryos for size by sieving through a 200 μm mesh screen as shown in *Figure 1(d)*. Large embryos respond the best to ABA and for best results, sieving through a 1000 μm screen is recommended. However, the larger the screen the fewer the embryos that are recovered. Embryos which pass through a 200 μm mesh screen are too

---

Protocol 2. *Continued*

small and underdeveloped and generally do not respond to the desiccation treatment (8).

10. Collect the size-selected embryos on damp filter paper, in 60 × 20 mm plastic Petri dishes, and transfer, without wrapping the Petri dish, to a desiccator with a controlled relative humidity of 97%. Proceed as outlined in *Protocol 1*, steps 11–13.

To rehydrate, aseptically transfer dried embryos to a filter paper on an agarose solidified, growth regulator-free, B5 medium containing 2% sucrose. Rapid rehydration of dried embryos, such as placing them directly into water, is detrimental to their survival. Culture embryos at 25 °C with a 16 hour photoperiod under fluorescent light.

4. Dried immature embryos of *Triticum aestivum*

Grow spring wheat, *Triticum aestivum* cv. Glenlea, under greenhouse conditions in 5 inch peat pots in a 1 : 1 : 1 soil : peat : sand mix under a 16 hour photoperiod of 20 °C day and 15 °C night temperatures. Plants should be watered daily and receive a 20 : 20 : 20 N : P : K treatment weekly.

Protocol 3. Production of dried immature wheat embryos

1. Collect the midsections of spikes, containing 15–20 developing seeds, 14 days post-anthesis and disinfect by dipping in 95% ethanol for 2 min followed by three washes with sterile distilled water.

2. Dissect out the embryos from developing seed under a dissecting microscope on an ethanol swabbed surface.

3. For dissection, peel back the surrounding leaf-like structures (glume, lemma, palea) from the developing grain and remove the smooth green grain.

4. Holding the grain, smooth side up, with aseptic forceps, and with the coleorhizal end towards the viewer, make a transverse incision with an aseptic scalpel across the grain at midpoint. Make a second incision along the side of the grain toward the coleorhizal end. Peel back the epidermal flap with the forceps and carefully lift out the white oval-shaped embryo with the tip of the scalpel.

5. Immediately transfer to a 60 × 20 mm plastic Petri dish containing agar solidified (0.8% Bacto agar) MS medium lacking growth regulators and modified to contain 5×10^{-5} M ABA. Culture embryos, five per dish, at 25 °C under a 16 h photoperiod.

6. Place cultured embryos on sterile filter paper in 60 × 20 mm plastic Petri dishes and transfer to a plastic desiccator as outlined in *Protocol 1*. We have found that embryos can be desiccated successfully if the embryos are started in desiccator No. 3 (RH 75%, *Table 2*).

To rehydrate, place dried embryos directly on filter paper on MS medium lacking growth regulators and modified to contain gibberellic acid (5×10^{-5} M GA$_3$). Assess germinating embryos for survival one week after being placed on the medium by monitoring for coleoptile and root growth.

References

1. Redenbaugh, K. (ed.) (1992). *Syn seeds: application of synthetic seeds to crop improvement*. CRC Press, Boca Raton, Florida.
2. Brown, D. C. W. (1988). *HortScience*, **23**, 530.
3. Murashige, T. and Skoog, F. (1962). *Physiol. Plant.*, **15**, 473.
4. Blaydes, D. F. (1966). *Physiol. Plant.*, **19**, 748.
5. Schenk, R. U. and Hildebrandt, A. C. (1972). *Can. J. Bot.*, **50**, 199.
6. Gamborg, O. L., Miller, R. A., and Ojima, K. (1968). *Exp. Cell Res.*, **50**, 151.
7. Huang, B. and Keller, W. A. (1989). *J. Tiss. Cult. Meth.*, **12**, 171.
8. Brown, D. C. W., Watson, E. M., and Pechan, P. M. (1993). *In Vitro Cell. Dev. Biol. Plant*, **29P**, 113.
9. Pechan, P. M., Bartels, D., Brown, D. C. W., and Schell, J. (1991). *Planta*, **184**, 161.
10. Winston, P. W. and Bates, D. H. (1960). *Ecology.*, **41**, 232.
11. Weast, R. C. (ed.) (1960). *CRC handbook of chemistry*, 57th edition CRC Press, Boca Raton, Florida, E-46.

7

Cryopreservation

ERICA E. BENSON

1. Introduction

Cryopreservation is the non-lethal storage of biological tissues at ultra-low temperatures. Protocols may be categorized by freezing method (controlled or ultra-rapid) and further defined by application (i.e to cells, meristems, embryos, callus). It is now possible to combine several cryopreservation techniques and apply them to a wide-range of plant systems. Thus, two different approaches are presented in this chapter; firstly, advice on the development of new protocols (e.g. for systems which have not been cryo-preserved previously) and secondly, examples of specific protocols, which are given to demonstrate the current diversity of cryopreservation methodology.

2. Theory and terminology

Cryopreservation comprises many steps, of which freezing is only one; suc-cessful recovery is dependent on the combined effects of cryogenics and pre- and post-freeze treatments.

2.1 Choice of plant material and pre-culture

Tissues must be selected from healthy plants and, in the case of *in vitro* material, culture parameters should be optimized before cryopreservation. Uniform suspensions composed of small groups of cytoplasmic, meristematic cells are more suitable than vacuolated cells which have a high water content (1). The duration of pre-freeze culture and light regimes may affect shoot-tip recovery (2,3). Similarly, developmental and physiological factors are import-ant determinants of recovery in embryo freezing.

2.2 Pre-growth

Within the context of this chapter, pre-growth will refer to treatments which do not protect plant tissues against exposure to liquid nitrogen, but which, in combination with other cryoprotective treatments, have the capacity to im-prove recovery. This definition is not categorical, and is given to aid protocol interpretation. Pre-growth may also involve cold hardening treatments or the application of additives, known to enhance plant stress tolerance (e.g. proline, trehalose, abscisic acid). Partial tissue dehydration can be achieved by the application of osmotically active compounds. The addition of relatively low (e.g. 1–5%) concentrations of dimethylsulfoxide (DMSO) during pre-growth often improves shoot-tip recovery.

2.3 Cryoprotective treatments

2.3.1 Chemical cryoprotection

Cryoprotectants are categorized as penetrating or non-penetrating. The former are thought to exert their protective effects through colligative action. Thus, the potential toxic effect of freeze-induced solute concentration is reduced. The protective action of non-penetrating cryoprotectants is largely by osmotic dehydration. Other cryoprotective properties include the ability to alter biochemical and structural properties of membranes and thereby enhance freeze tolerance.

2.3.2 Vitrification

Avoidance of ice formation in biological tissues exposed to low and ultra-low temperatures reduces damage. This can be achieved through vitrification, a process in which ice formation cannot take place because the aqueous solution is too concentrated to permit ice crystal nucleation. Instead, water solidifies into an amorphous 'glassy' state. In cryopreservation, vitrification usually involves the application of a highly concentrated solution of osmotically active compounds, many of which are used as 'traditional' cryoprotectants or pre-growth additives at lower concentrations. Although vitrification circumvents ice formation, the amorphous state is metastable and the transfer of vitrified materials between different temperatures (e.g. during warming) can lead to crystallization. Care must be taken in the application of vitrification 'cocktails' which become toxic on prolonged exposure. Furthermore, damaging osmotic affects may also occur on their removal.

2.3.3 Cryoprotective dehydration

If cells are sufficiently dehydrated they may be able to withstand immersion in liquid nitrogen without the further application of traditional cryoprotectant mixtures. Dehydration can be achieved by growth in the presence of high concentrations of osmotically active compounds (sugars, polyols) and/or air desiccation in a sterile flow cabinet or over silica gel. Dehydration reduces the amount of water available for ice formation. If sufficiently dehydrated, cellular solute concentrations are high enough to promote vitrification.

2.3.4 Encapsulation and dehydration

This involves the encapsulation of tissues in calcium alginate beads which are pre-grown in liquid culture media containing high concentrations of sucrose.

The beads are transferred to a sterile air flow in a laminar cabinet and desiccated further. After these treatments the tissues are able to withstand exposure to liquid nitrogen without the application of chemical cryoprotectants (4).

2.4 Freezing and storage

The means by which tissues are transferred from growth temperatures to ultra-low temperatures can be an important factor in post-freeze survival. Samples may be directly plunged into liquid nitrogen (rapid freezing), or progressively chilled at a controlled rate to an intermediate temperature. On reaching this temperature they may be transferred directly to liquid nitrogen, or held for a set time before transfer. In some cases it may be necessary to control the point at which external ice occurs. This event is termed ice nucleation or seeding and describes the process in which a molecule, particle, or vessel wall provides a surface for crystal growth. Water can be supercooled below 0 °C; the degree of supercooling is dependent upon the prevention of ice nucleation. The state of phase transition produces an exotherm (thermal heat of crystallization) evidenced as a few degrees increase in temperature. Without external intervention the initiation of ice crystals is a relatively random process and this is termed heterogeneous nucleation. However, it is possible to initiate external ice nucleation by rapidly touching the outside of the vessel with chilled instruments or by agitation. On external ice nucleation, there exists a vapour pressure deficit between the inside and the outside of the cell. Water leaves the cell and, as a result, protective dehydration occurs. Thus, the ice nucleation characteristics of a system may be an important factor in post-freeze survival. Samples are maintained in storage dewars and levels of liquid nitrogen must be regularly monitored and replenished as required.

2.5 Thawing

The temperature and rate at which tissues are thawed is dependent on the freezing method. Thawing is particularly critical in vitrified tissues as ice crystallization can occur during re-warming. In conventional freezing methods, thawing is usually achieved by plunging the cryovials into sterile water maintained at 40–45 °C. Once all the ice has melted the samples are removed. Tissues which have been frozen by encapsulation and/or dehydration are frequently thawed at ambient temperatures. Vitrified and excessively dehydrated tissues may require a sequential lowering of media osmotica during thawing and early recovery.

2.6 Recovery

Post-thaw recovery comprises several stages: assessments of viability and cell division, morphological changes (leaf expansion, callusing, embryo development), regeneration of shoots and plants, and resumption of metabolite production. In the case of organized structures, recovery is frequently induced by the application of hormones.

3. Cryopreservation equipment

Simple freezing chambers utilizing a cooled solvent system may be used. However, a typical, modern programmable freezer comprises a liquid nitrogen cooled sample chamber, to which nitrogen is pumped from an accessory dewar or storage tank. Cooling rates, holding times, and terminal transfer temperatures are programmed, and chamber and sample temperatures are monitored by thermocouples. Models can be purchased from several companies (Planar UK, Cryomed USA, L'Air Liquide, France).

You will need the following equipment:

- a reliable source of liquid nitrogen
- safety equipment (gloves, apron, face shield, pumps for dispensing liquid nitrogen from and to large storage dewars, trolleys for the transport of dewars)
- small (1–2 litre) liquid nitrogen resistant dewar(s)
- dewar(s) for the routine storage of liquid nitrogen
- dewar(s) for the long-term storage of specimens—these may be equipped with an alarm system which is activated if the liquid nitrogen level falls below a critical level
- cryovials, straws, boxes, canes, racks
- a domestic refrigerator (−20 °C)
- a programmable freezer with dewar and pump
- a water-bath for thawing at 40–50 °C

4. A cryopreservation strategy

Before applying cryopreservation to a new system it is essential to have a strategy which will progressively examine the effects of different parameters. Thus, as well as investigating a routine method, it may be necessary to develop new methods or optimize standard protocols for specific plant species and/or tissue types. Some tissues or species may be particularly sensitive to a cryoprotectant or to the prolonged exposure to a vitrification 'cocktail'. Similarly, tissues are only able to withstand a certain level of cryoprotective dehydration. Thus, it is important to perform 'unfrozen control' experiments for additives or procedures used in pre-growth, dehydration, cryoprotection, vitrification, encapsulation, and recovery. Once any possible deleterious additive or pre-growth effects have been discounted, it is possible to proceed to the freezing part of the method. Two options may be considered, rapid freezing or controlled rate cooling. Choice will be dependent upon the protocol and tissue system, whether it is possible to freeze directly or by controlled cooling, and the availability of specialized freezing equipment. A strategy for developing a controlled cooling protocol would, for example, involve testing cooling rates and transfer temperatures. Tissue recovery after cryopreservation is related to *in vitro* manipulations and/or the physiological and developmental status of the tissues. These factors are critical, and their importance in determining post-thaw survival cannot be over-stressed.

5. Protocols for method development

5.1 Pre-growth

The following protocol offers an example which may be used for a range of different tissues. Treatments can also include cold-hardening strategies which are frequently linked to seasonal factors such as dormancy (5). Because of the specific nature of cold-hardening treatments it is not possible to present a general methodology.

Protocol 1. Optimization of pre-growth regimes

1. Prepare standard liquid or semi-solid medium and add one or more of mannitol, sucrose, proline, sorbitol, trehalose, or glucose[a] in a range of 0.3–0.5 M.
2. Culture tissues with the additives for one to five days.
3. At daily intervals, return tissues to standard growth.
4. Determine survival for each treatment.

[a] Examples are from a study by Goldner *et al.* (6).

In developing a pre-growth method for shoot-tips it is essential to standardize the donor material as variation in response to freezing is a major problem in shoot-tip cryopreservation (2). Dissection injury can often be a significant determinant of post-freeze survival, particularly if the shoot-tips are frozen directly after excision (7).

Protocol 2. Shoot-tip dissection

1. Select apical shoots at the same developmental age or cultures of the same subculture age. For rapidly growing cultures, nodal sections can be cultured for five to ten days (dependent on species) and the developing axillary bud used for cryopreservation.

Protocol 2. *Continued*

2. Remove the shoot-tip and place on sterile filter papers under a binocular microscope. With the aid of two fine-gauge hypodermic needles (fitted to a syringe), gently dissect away any fully expanded leaves and cut the shoot-tip to a size of 0.3–0.5 mm (dependent on species). Retain three or four non-expanded leaf primorida on each shoot-tip. It may be necessary to optimize both shoot-tip size and the period required for recovery.

3. Immediately after dissection place the shoot-tip on to a filter paper bridge moistened with standard liquid culture medium. The application of 0.5–5% (v/v) DMSO at this stage may enhance survival.

4. Pre-grow for one to three days before cryopreservation.

5.2 Chemical cryoprotection

There is a wide-range of cryoprotectants (8,9). *Protocol 3* describes two routine mixtures and procedures for assessing their optimum rate and temperature of application (1).

Protocol 3. Cryoprotection of cell and callus cultures

1. Prepare one of the following cryoprotectants in standard liquid culture medium:
   - 1 M DMSO + 1 M glycerol + 2 M L-proline
   - 1 M DMSO + 1 M glycerol + 2 M sucrose

2. Divide the cryoprotectant into two aliquots, chill one aliquot in iced water and maintain the second at 25 °C.

3. Remove excess liquid from replicate flasks of cell suspension cultures leaving a known volume of culture.

4. Add chilled or 25 °C cryoprotectant mixture to replicate cultures in the volume ratio of 1:1, by either direct or gradual addition (2–3 ml aliquots over 30 min). Maintain the chilled replicate on ice throughout the incubation time.

5. Cryoprotect for 1 h (inclusive of addition time).

6. Empty flask contents on to filter papers, assess cell viability in a few cells (*Protocols 20* and *21*). Transfer remaining cells to semi-solid culture medium and assess re-growth. Repeat the process with the second cryoprotectant and determine the optimum procedure.

Protocol 4 is a general procedure for developing a cryoprotection strategy for meristem structures (2,3,10,11).

Protocol 4. Cryoprotection of meristem structures

1. Prepare cryoprotectant solutions of DMSO in standard liquid medium in the range 5–15% (v/v).

2. Add 10–15 meristem structures (shoot or root-tips) to 0.5–0.75 ml of cryoprotectant medium (at 25 °C).[a]

3. Cryoprotect for 15 min to 1 h.

4. Remove meristem structures, blot dry on filter paper, transfer to recovery medium, and assess re-growth in comparison to uncryoprotected controls. Select the least damaging treatment for cryopreservation.

[a] For sensitive systems (11), test the effects of addition rate and temperature *Protocol 3*).

5.3 Cryoprotective dehydration treatments

Tissues will only tolerate a certain degree of dehydration. The lethal minimum level of tissue water content can be species-specific and must be determined before embarking on any freezing regime. *Protocols 5* and *6* can be used singularly or in combination (usually consecutively). Cryoprotective dehydration treatments are suitable for zygotic embryos (12), somatic embryos (13,14,15), embryonic axes (16), nodal stem segments (17), and calli (18).

Protocol 5. Cryoprotective chemical dehydration

1. Prepare standard liquid or solid medium and add one of the following: mannitol, sucrose, glucose, or sorbitol, in the concentration range 0.50–1.5 M.

2. Determine tissue fresh weight then transfer to media.

3. Dehydrate tissues for short (2–24 h) and long-term periods (one to seven days).

4. Determine fresh weights and return to standard medium. Determine dry weights of replicate samples.

5. Correlate maximum water loss with minimum mortality. Select this treatment for pre-freeze dehydration.

Protocol 6. Cryoprotective air dehydration

1. Determine tissue fresh weight.

Protocol 6. *Continued*

2. Dehydrate tissues for several time intervals in the range 1–16 h by using one of the following methods:
   - sterile air flow
   - flash drying in a pressured stream of sterile air
   - drying over silica gel

3. Determine fresh weights of tissues and return to standard culture medium. Determine dry weights of replicate samples.

4. Correlate maximum water loss with minimum mortality.

5.4 Encapsulation and dehydration

This procedure must be developed in conjunction with cryoprotective dehydration treatments. At each stage, determine bead dehydration by assessments of fresh and dry weights (*Protocols 5* and *6*). *Protocol 7* was developed by Fabre and Dereuddre for potato shoot-tips (4).

Protocol 7. Cryoprotective encapsulation and dehydration

1. Prepare:
   - calcium-free standard liquid medium containing 3% (w/v) sodium alginate (2% viscosity Kelp Na$^+$ salt)[a]
   - standard liquid medium containing 100 mM $CaCl_2$

2. Dissect shoot-tips (see *Protocol 2*).

3. Add shoot-tips to a flask and add ∼ 25 ml alginate solution, then swirl until shoot-tips are completely submerged. Take care not to form air bubbles.

4. Place ∼ 50 ml of 100 mM $CaCl_2$ medium into a 100 ml conical flask. Using a 5 ml 'Pipetteman' withdraw 3 ml of the alginate solution and, with gentle manipulation, take up several shoot-tips at the same time.

5. Wipe away any excess alginate from the tip (failure to do so will result in asymmetric beads).

6. Place the pipette tip over the flask containing the calcium medium and dispense drops into the liquid; beads form as the alginate solidifies on contact with the Ca^{2+} ions. It is possible to encapsulate one to three shoot-tips per bead.

7. Pour off excess medium, pick out the beads with the shoot-tips, and allow the beads to polymerize for 30 min.

8. Divide the encapsulated shoot-tips into replicate samples, blot dry on filter paper, and transfer to the following:
   - standard semi-solid culture medium[b]

- A range of cryoprotective dehydration media made up in liquid culture medium (see *Protocol 5*)

9. Dehydrate the beads in liquid culture for short (2–16 h) and long-term periods (one to five days).

10. At appropriate time intervals remove replicate samples from each treatment; place one replicate on to standard semi-solid medium.

11. Transfer the remaining replicates to open Petri dishes in a sterile laminar air flow (see *Protocol 6*), air dry for 1–4 h, and at each time interval transfer to standard semi-solid medium.

12. Determine levels of survival for each stage and select the treatment which gives maximum survival for maximum dehydration.

[a] Place the liquid medium on a magnetic-stirrer and stir vigorously while adding small amounts of the alginate. Discard solutions which have flocculated.
[b] In some cases shoot-tips entrapped in alginate are unable to proliferate and shoot-tip removal is necessary (19).

5.5 Vitrification

Vitrification solutions are potentially very toxic, and critical factors include the cold acclimation of tissues before vitrification, the temperature of application of vitrification solution, and the duration of exposure. There have been several successful reports of vitrification (9,20,21). Sakai *et al.* (22) have developed the 'PVS2' vitrification solution which has proved applicable to a range of systems (18,22,23).

Protocol 8. Application of cryoprotective 'PVS2' vitrification solution

1. Prepare the following solutions:
   - 'PVS2' stock solution: 30% (w/v) glycerol + 15% (w/v) ethylene glycol + 15% (w/v) DMSO in standard liquid medium containing 0.15 M sucrose
   - 60% (v/v) 'PVS2' solution in standard liquid medium containing 0.4 M sucrose
   - 'unloading' solution: 1.2 M sucrose in standard liquid medium

2. Add 1.0 ml 60% (v/v) 'PVS2' solution to the tissues.

3. Incubate at 25 °C for 5–20 min.

4. Remove 60% solution, replace with 1 ml chilled 100% 'PVS2' and incubate tissues in an iced water-bath for 3–10 min.

5. Transfer to 25 °C and remove excess 'PVS2' solution.

6. Add 1.0 ml unloading solution and change three times. Incubate in unloading solution for 5–30 min.

Protocol 8. *Continued*

7. Transfer to filter papers overlaid on standard semi-solid culture medium. On the following day transfer to fresh medium. Evaluate survival for each treatment.

5.6 Freezing

In the case of cell suspension cultures, it is often possible to apply the method of Withers and King (1). However, shoot-tips often have highly species-specific controlled cooling requirements (24). *Protocol 9* offers one possible regime for optimizing controlled cooling, although this is by no means definitive. In the case of freeze-recalcitrant tissues the lengthy development of a controlled cooling method must be offset by the possibility of using a different approach (e.g. vitrification, encapsulation). If cultures prove consistently difficult to freeze by controlled cooling it may be necessary to consider the more complex interactive effects of cryoprotection and freezing (25).

Protocol 9. The development of controlled cooling methods for cells and organized structures

1. Transfer cryoprotected tissues to a number of replicate cryovials.
2. Set the starting temperature of the freezer to 0 °C or 25 °C, depending on the temperature of cryoprotection (see *Protocols 3* and *4*).
3. Programme one of the following cooling rates: -5, -1.0, -0.5, -0.25, -0.1 °C/min.
4. Cool to intermediate transfer temperatures: -0, -10, -15, -20, -25, -30, -40 °C, and remove vial(s) as appropriate.
5. Thaw at 45 °C until the ice has melted and transfer the tissues to standard semi-solid medium.
6. Determine optimum transfer temperature. Repeat under optimum conditions, but this time transfer the tissues to liquid nitrogen.
7. Assess recovery. If the programme was unsuccessful, repeat steps **1** to **6** using different cooling rates. Use 'control' samples which have not been exposed to liquid nitrogen to indicate intermediate survival.
8. If tissues are still not able to survive exposure to liquid nitrogen test the following variables:
   - 'seeding' ice crystallization (*Protocol 11*)[a]
   - holding for 20–40 min at the terminal temperature
   - stepwise cooling (e.g. -1.0 °C/min to -15 °C, followed by -0.5 °C/min to -40 °C)

[a] Whilst seeding has been used to advantage in several systems it is difficult to control precisely and this may result in variable recovery responses.

5.7 Thawing and recovery

There are few variables at the thawing stage of method development, and examples are given in the following section. Recovery regimes are species-specific and influenced by tissue culture methods. A general strategy for the development of recovery protocols is not given.

6. Protocols for specific applications

6.1 Chemical cryoprotection and controlled cooling

Protocol 10 is based on the method of Withers and King (1), with minor modifications added to the recovery procedure (26). This protocol has proved generally successful, but may require considerable modification for difficult systems (27).

Protocol 10. Cryopreservation of cell cultures

1. Select cell culture, taking into consideration pre-growth factors (see *Protocol 1*).

2. Remove excess culture medium to a known volume and add the ice-chilled cryoprotectant (1.0 M DMSO + 1.0 M glycerol + 2.0 M sucrose) in a volume ratio of 1:1.

3. Mix the cells thoroughly, cryoprotect on ice for 1 h, and transfer 1–1.75 ml aliquots of the cryoprotectant–cell mixture to 2.0 ml cryo-vials. Load the vials on to a cane and transfer to a programmable freezer, pre-chilled to 0 °C.

4. Use the following conditions to initiate the chilling programme:
   - Chilling rate −1 °C/min
   - Terminal temperature −35 °C
   - Holding time 30 min

5. On completion, quickly transfer the canes to a long-term storage dewar containing liquid nitrogen.

6. Transfer vials to a 45 °C water-bath, and, once the ice has melted, pour their contents on to several layers of 5.0 cm filter papers in a 9.0 cm Petri dish.

7. Transfer the cells and the top few layers of filter papers to semi-solid culture medium.

8. Check early post-thaw viability (see *Protocols 20* and *21*).

9. After one day, transfer the uppermost layers of filter papers/cells to fresh semi-solid medium and continue to transfer on a daily basis (one to four days). Transfer cells to fresh medium without filters (five to six days).

Protocol 10. *Continued*

10. Initiate liquid suspensions when re-growth is sufficient to permit a normal subculture inoculation.

Protocol 11 describes shoot-tip preparation (2,3) followed by a controlled cooling method developed by Towill (28,29) for potato shoot-tips. This example is cited as both ice 'seeding' and stepwise programmable freezing are included. However, in common with most controlled freezing techniques, it is not generally applicable to shoot-tips from a wide-range of species and would require species-specific development (see *Protocol 9*). The medium described by Towill (28,29) for potato shoot-tip recovery has also proved successful for other systems (Benson, unpublished observations).

Protocol 11. Cryopreservation of shoot-tips

1. Dissect shoot-tips (see *Protocol 2*) and pre-grow for 24 h on filter paper bridges with liquid culture medium.

2. Transfer batches of 15–20 shoot-tips to cryovials containing 0.5–1.0 ml of 10% (v/v) DMSO.

3. Cryoprotect for 1 h at 25 °C.

4. Transfer to a programmable freezer pre-chilled to −5 °C and chill for 10 min.

5. Induce ice ('seeding') by mechanical agitation or by touching the outside of the vial with liquid nitrogen-chilled forceps.

6. Chill for a further 10 min at −5 °C.

7. Cool at a rate of −0.4 °C/min to a terminal temperature of −35 °C and transfer to liquid nitrogen.

8. Thaw at 45 °C.

9. Empty the contents on to filter papers to remove cryoprotectant.

10. Transfer shoot-tips to standard semi-solid medium containing the following hormone combination:
    - 0.2 mg/litre gibberellic acid (GA_3)
    - 0.5 mg/litre indole acetic acid
    - 0.5 mg/litre zeatin (mixed isomers)

11. When shoot regeneration has occurred transfer to standard culture medium.

Transformed 'hairy root' cultures are also amenable to cryopreservation using controlled chilling to 0 °C before direct transfer to liquid nitrogen (10).

Protocol 12. Cryopreservation of transformed root cultures

1. Excise 2–4 mm root-tip sections with the aid of hypodermic needles and a binocular microscope. Perform all manipulations under liquid culture medium as the roots readily desiccate, making dissection difficult.

2. Transfer root-tips to 2 ml cryovials and add 0.5 ml of 10% (w/v) DMSO (in liquid culture medium).

3. Cryoprotect for 1 h at 25 °C.

4. Transfer the vials to a programmable freezer and chill from 25 °C to 0 °C at a rate of −1.0 °C/min.

5. Transfer immediately to liquid nitrogen.

6. Thaw root-tips at 45 °C then empty the vial contents on to several layers of filter paper. Drain excess cryoprotectant and transfer to semi-solid culture medium.

Cryopreservation using pre-growth and cryoprotective treatments, in conjunction with controlled cooling, also offers a potential method of conserving immature embryos of recalcitrant seed species (30). Controlled cooling methods are dependent upon the use of expensive programmable freezing equipment. Lecouteux *et al.* (31) report a simplified approach for cryopreserving somatic embryos (*Protocol 13*).

Protocol 13. Simplified controlled freezing of somatic embryos

1. Pre-grow somatic embryos (at heart and torpedo stage) for 24 h in standard liquid medium containing 0.4 M sucrose (at a density of 100 embryos/10 ml in shaker culture).

2. Transfer 100 embryos to a cryovial (Nunc 368632) containing 1.8 ml of the cryoprotectant growth medium.

3. Place the vials in a domestic −20 °C freezer for 24 h.

4. Transfer directly from −20 °C to liquid nitrogen.

5. Thaw at 40 °C then transfer to standard medium containing 0.015 M sucrose.

6.2 Vitrification methods

'PVS2' solution (see *Protocol 8*) was developed by Yamada *et al.* (23) for the cryopreservation of white clover shoot-tips. The cryopreservation of shoot-tips using controlled cooling is often variable, and vitrification is a promising alternative approach. By omitting the pre-growth stages and optimizing vitrification 'cocktail' application, 'PVS2' may also be applied to cell suspension cultures (18, 22).

Protocol 14. A vitrification procedure for shoot-tips

1. Excise shoot-tips (see *Protocol 2*).
2. Pre-grow at 4 °C[a] in standard semi-solid medium containing 1.2 M sorbitol for two days.
3. Transfer shoot-tips to 2 ml cryovials, apply 'PVS2' vitrification solution (see *Protocol 8*) by either of the following procedures:
   - incubate in 'PVS2' solution for 5 min at 25 °C
   - incubate in 'PVS2' solution for 15 min at 0 °C
4. Directly plunge into liquid nitrogen.
5. Warm at 25 °C.
6. Remove the vitrification solution and wash the shoot-tips twice with 1.2 M sucrose solution (see *Protocol 8*).
7. Transfer shoot-tips to a 5.0 cm filter paper overlaying standard semi-solid medium. Transfer daily to fresh filter papers until regeneration is evident, and then transfer to standard growth conditions.

[a] For chill sensitive species it is possible to pre-grow at higher temperatures.

6.3 Cryoprotective dehydration methods

Cryopreservation of nodal stem segments using cryoprotective dehydration offers a simple and useful alternative for the storage of 'shoot' material (17). A major advantage of the technique is that it does not require time-consuming dissections. However, its wider applicability requires further investigation. It is very important to standardize stem segment size as this can greatly affect the dehydration properties of the material (Benson, unpublished observations). The following protocol was developed for axillary buds of Asparagus by Uragami *et al.* (17)

Protocol 15. A dehydration method for cryopreserving nodal stem sections

1. Remove uniformly sized (5 mm) stem sections from entire plants.
2. Place on semi-solid medium containing 0.7 M sucrose and incubate for two days under standard conditions.

---

3. Transfer nodes to a nylon membrane contained in a dish with 15 g dry silica, seal with Parafilm, and desiccate for 4–8 h.

4. Transfer nodes to cryovials and plunge directly into liquid nitrogen.

5. Thaw at 25 °C and return to standard growth medium.

Protocol 16 describes a simple 'air drying' method, developed for the cryopreservation of embryonic axes of tea (16). A similar approach has been used for zygotic embryos of *Musa* spp. (12). Desiccation to a moisture content of 13% was optimum for tea; the method may require further development (see *Protocol 6*) for other species.

Protocol 16. Cryopreservation of zygotic embryonic axes by air dehydration

1. Aseptically excise embryonic axes from the seeds.

2. Desiccate axes in sterile air flow for 3 h.

3. Transfer to a cryovial and immerse directly in liquid nitrogen.

4. Thaw in a water-bath at 37–38 °C.

5. Place the axes on water-moistened filter paper (without additives) for up to ten days.

6. Transfer to standard nutrient medium to permit continued development.

Assy-Bah and Engelmann (32,33) have recently developed cryoprotective dehydration techniques for the cryopreservation of immature and mature embryos of seed-recalcitrant coconut (*Protocol 17*). Application to other systems would require the determination of precise dehydration parameters (see *Protocols 5* and *6*) and species-specific factors related to embryo development.

Protocol 17. Cryopreservation of immature and mature zygotic embryos using combined cryoprotective dehydration

A. *Immature embryos*

1. Place embryos on to standard semi-solid medium containing 600 g/litre glucose (replacing standard sugars) and 15% (v/v) glycerol.

2. Dehydrate for 4 h, transfer to cryovials, and plunge directly into liquid nitrogen.

3. Thaw for 30 sec at 40 °C and transfer to standard medium for recovery.

161

Protocol 17. *Continued*

B. *Mature embryos*

1. Place embryos in open Petri dishes (without medium) and desiccate in a sterile air flow for 4 h.

2. Transfer to medium containing 600 g/litre glucose and 15% (v/v) glycerol, dehydrate for 15–24 h (cultivar dependent).

3. Freeze, thaw, and recover as for immature embryos.

A number of different approaches are available for cryopreserving somatic embryos (31,34,35). The following cryoprotective dehydration method was originally developed for somatic embryos of oil palm. A major advantage is that it does not require the use of cryoprotectants (13,14).

Protocol 18. The cryopreservation of somatic embryos using cryoprotective dehydration

1. Culture embryonic masses on standard semi-solid medium containing 0.3 M sucrose for two months.

2. Transfer to standard semi-solid medium containing 0.75 M sucrose for seven days.

3. Dissect out embryonic tissue, transfer to cryovials, and plunge directly into liquid nitrogen.[a]

4. Thaw at 45 °C then transfer to standard semi-solid medium containing 0.3 M sucrose. After two weeks transfer to standard medium containing 0.1 M sucrose.

[a] For some systems it may be necessary to use controlled cooling (see *Protocol 9*).

6.4 Encapsulation and dehydration

Protocol 19 is based on the encapsulation technique initially established for *Solanum phureja* shoot-tips (4). With modifications, it has also been applied to carrot somatic embryos (36), shoot-tips of grape (37), pear (19), mulberry (38), and microspore embryos of oil seed rape (39). This protocol combines methodology for cryopreserving shoot-tips (4) and somatic embryos (36, 40).

Protocol 19. Cryopreservation of shoot-tips and somatic embryos by encapsulation and dehydration

1. Dissect shoot-tips (see *Protocol 2*).

2. Encapsulate shoot-tips or torpedo stage somatic embryos as described in *Protocol 7*.

3. Transfer beads to standard liquid medium containing 0.75 M sucrose (for shoot-tips) or 0.3 M sucrose (for embryos). Place on a shaker and dehydrate for 72 h (shoot-tips) or 18 h (embryos).

4. Transfer beads to filter papers, blot dry, remove filters, and desiccate in a sterile air flow for 4 h.

5. Transfer the dried beads to a cryovial and plunge directly in liquid nitrogen.[a]

6. Thaw at ambient temperature (~25 °C) and transfer the beads to semi-solid recovery medium. If necessary excise tissues from the beads to aid recovery.

[a] For potato and possibly other systems it may be necessary to use controlled cooling (*Protocol 9*).

7. Post-thaw viability, recovery, and stability

Vital staining techniques are useful in determining early post-thaw survival. However, they are not always predictive of long-term viability and recovery and their use is cautioned as the sole means of assessing survival. Ideally, tests should be performed at different stages after thawing. The only definitive indication of recovery is re-growth.

7.1 Fluorescein diacetate vital stain

The fluorescein diacetate (FDA) vital stain is dependent upon the ability of esterases in viable cells to cleave the stain, which then fluoresces yellow/green under a UV microscope (41).

Protocol 20. Fluorescein diacetate (FDA) vital stain

1. Make up a stock solution of 0.1% (w/v) FDA in acetone. This may be stored at 4 °C. Add a few drops of stock to 10 ml of liquid culture medium.

2. Add a few drops of FDA stain to cells on a microscope slide. Allow the stain to penetrate for a few minutes, place a cover-slip over the cells, and examine under a UV fluorescence microscope using a blue/violet filter.

3. Count the number of fluorescing cells in a dark field of view and count the total number of cells in the same field under bright illumination.

4. Repeat for at least five fields of view and for each calculate:

$$\% \text{ viability} = \frac{\text{number of fluorescent cells}}{\text{total number of cells}} \times 100.$$

5. Determine the mean of the results, and express as a percentage of the total frozen cell population or of the population of unfrozen control cells.

7.2 The triphenyl tetrazolium chloride (TTC) assay

This is a useful alternative for larger specimens (42). Dehydrogenase activity of viable cells reduces TTC to a red formazan product which is measured spectrophotometrically.

Protocol 21. The triphenyl tetrazolium chloride (TTC) test

Reagents

- 0.6% (w/v) TTC in 0.05 M Na_2HPO_4/KH_2PO_4 buffer pH 7.4 + 0.05% (v/v) Tween 80
- 95% (v/v) ethanol

Method

1. Place 50–100 mg (fresh weight) plant tissue in a test-tube, add 3 ml of TTC reagent and infiltrate under vacuum for 10–20 min.

2. Incubate at 28 °C overnight.

3. Add 7 ml 95% ethanol and extract the coloured complex by boiling in a water-bath for 4 min.

4. Mix the samples thoroughly and centrifuge at 500–1000 g to precipitate the cell debris.

5. Measure the absorbance of the supernatant at 490 nm.

6. Express viability as absorbance$_{490nm}$/g fresh weight or as a percentage of the value for unfrozen controls.

Further assessments of recovery are usually specific to the system under study and it is therefore outside the scope of the chapter to present detailed methodologies. However, examples include monitoring the resumption of post-thaw secondary metabolite activity (43), the retention of genetically manipulated characteristics (10), and the ability to form protoplasts (26). Studies of cryoinjury may be important in developing protocols for freeze-recalcitrant systems. A range of assays have been developed to test for oxidative injury during cryopreservation (44,45). Non-destructive methods for assessing freezing damage may be especially useful. Thus, ethylene may be used as a marker of freezing and chilling stress (46,47), chemiluminescence as a marker of oxidative damage (48), and infra-red spectroscopy as an indicator of viability (49). Retention of genetic fidelity in plants recovered from cryopreservation is vital. Molecular probes for ribosomal DNA may be used to study restriction fragment length polymorphisms in plants recovered from cryopreservation. This offers a potentially important method of assessing genetic stability (50). Similarly, the use of flow cytometry to assess post-freeze ploidy stability may prove a rapid and valuable technique (51).

8. Conclusions

Progress in the area of plant cell, tissue, and organ cryopreservation has been considerable since the first edition of this volume. The freezing method developed by Withers and King (1) for the cryopreservation of plant cells is now in routine use. Adaptations of this method may also be applied to callus tissues. Vitrification is proving a possible alternative to traditional cryoprotectants and the cryopreservation of somatic embryos is also possible using a range of techniques. Cryopreservation is becoming increasingly important for the conservation of recalcitrant seed species. Whilst there still remain problems with the cryopreservation of shoot-tips from several species, new techniques involving encapsulation offer promising alternatives to 'traditional' methods. Within the constraints of this chapter it has not been possible to present every aspect of method development. 'Traditional' techniques have been included for their proven routine applicability or their continued importance in protocol development. However, major emphasis has been given to newer techniques which hopefully have the potential to become routine protocols of the future. A significant trend, presently emerging from the field of plant cryopreservation, concerns the development of technologically simple methods and their wider application to several different types of cells and organs. The ability to cryopreserve plant tissues without the need for specialized freezing equipment has the immediate benefit of making cryopreservation accessible to many laboratories. This will greatly assist the continued progress of plant tissue cryopreservation and most importantly, increases the possibility of developing a wider range of routine cryopreservation protocols. Finally, as cryopreservation becomes an important aspect of genetic resource conservation it is important to underpin the practical aspects of method development with a greater understanding of basic plant cryobiology. Thus, further progress within the application of cryopreservation techniques will be aided by basic studies concerning freezing and dehydration injury, cryoprotectant mode of action, and post-thaw regeneration. As the maintenance of genetic stability is fundamental to the safe storage of genetic resources, continued investigations of the molecular aspects of post-thaw stability are of priority. These may be of particular significance for the increased application of cryopreservation to genetically engineered plant cell and organ cultures.

Acknowledgements

The writing of this chapter was aided by the experience I gained as a visiting scientist in the Laboratoire de Ressources Genetiques, ORSTOM, Montpellier, France. I am most grateful to all my ORSTOM colleagues for their kind assistance, and especially to Dr Florent Engelmann, Dr Beatrice Assy-

Erica E. Benson

Bah, Dr Jaques Fabre, and Nathalie Chabrillange for their helpful and enthusiastic cryopreservation discussions.

References

1. Withers, L. A. and King, P. J. (1980). *Cryo-Lett.*, **1**, 213.
2. Benson, E. E., Harding, K., and Smith, H. (1989). *Cryo-Lett.*, **10**, 323.
3. Harding, K., Benson, E. E., and Smith, H. (1991). *Cryo-Lett.*, **12**, 17.
4. Fabre, J. and Dereuddre, J. (1990). *Cryo-Lett.*, **11**, 413.
5. Tyler, N., Stushnoff, C., and Gusta, L. V. (1988). *Can. J. Plant Sci.*, **68**, 1169.
6. Goldner, E. M., Seitz, U., and Reinhard, E. (1991). *Plant Cell Tiss. Organ Cult.*, **24**, 19.
7. Henshaw, G. G., O'Hara, J. F., and Stamp, J. A. (1985). In *Cryopreservation of plant cells and organs* (ed. K. K. Kartha), pp. 159–70. CRC Press, Boca Raton, Florida.
8. Finkle, B. J., Zavala, M. E., and Ulrich, J. M. (1985). In *Cryopreservation of plant cells and organs* (ed. K. K. Kartha), pp. 76–113. CRC Press, Boca Raton, Florida.
9. Langis, R. and Steponkus, P. L. (1990). *Plant Physiol.*, **92**, 666.
10. Benson, E. E. and Hamill, J. D. (1991). *Plant Cell Tiss. Organ Cult.*, **24**, 161.
11. Demeulemeester, M. A. C., Panis, B. J., and De Proft, M. P. (1992). *Cryo-Lett.*, **13**, 165.
12. Abdelnour-Esquivel, A., Mora, A., and Villalobos, V. (1992). *Cryo-Lett.*, **13**, 159.
13. Engelmann, F. (1990). *C. R. Acad. Sci. Paris*, S. III, **311**, 679.
14. Engelmann, F., Duval, Y., and Dereuddre, J. (1985). *C. R. Acad. Sci. Paris*, S. III, **301**, 111.
15. Dumet, D., Engelmann, F., Chabrillange, N., and Duval, Y. (1993). *Plant Cell Rep.*, **12**, 352.
16. Chaudhury, R., Radhamani, J., and Chandel, K. P. S. (1991). *Cryo-Lett.*, **12**, 31.
17. Uragami, A., Sakai, A., and Nagai, M. (1990). *Plant Cell Rep.*, **9**, 328.
18. Uragami, A. (1991). *Res. Bull. Hokkaido Natl. Agric.*, **156**, 1.
19. Dereuddre, J., Scottez, C., Arnaud, Y., and Duron, M. (1990). *C. R. Acad. Sci. Paris*, S. III, **310**, 317.
20. Towill, L. E. (1990). *Plant Cell Rep.*, **9**, 178.
21. Langis, R. L. and Steponkus, P. L. (1991). *Cryo-Lett.*, **12**, 107.
22. Sakai, A., Kobayashi, S., and Oiyama, I. (1990). *Plant Cell Rep.*, **9**, 30.
23. Yamada, T., Sakai, A., Matsumura, T., and Higuchi, S. (1991). *Plant Sci.*, **78**, 81.
24. Kartha, K. K. (1985). In *Cryopreservation of plant cells and organs* (ed. K. K. Kartha), pp. 115–34. CRC Press, Boca Raton, Florida.
25. Heszky, L. E., Jekkel, Z., and Abdel-Hamid, A. (1990). *Plant Cell Tiss. Organ Cult.*, **21**, 217.
26. Meijer, E. G. M., van Iren, F., Scrinjnemakers, E., Hensgens, L. A. M., van Zijderveld, M., and Schilperoort, R. A. (1991). *Plant Cell Rep.*, **10**, 171.
27. Panis, B. J., Withers, L. A., and De Langhe, E. A. L. (1990). *Cryo-Lett.*, **11**, 337.
28. Towill, L. E. (1983). *Cryobiology*, **20**, 567.
29. Towill, L. E. (1984). *Cryobiology*, **5**, 319.

30. Pence, V. C. (1991). *Plant Cell Rep.*, **10**, 144.
31. Lecouteux, G., Florin, B., Tessereau, H., Bollon, H., and Petiard, V. (1991). *Cryo-Lett.*, **12**, 319.
32. Assy-Bah, B. and Engelmann, F. (1992). *Cryo-Lett.*, **13**, 67.
33. Assy-Bah, B. and Engelmann, F. (1992). *Cryo-Lett.*, **13**, 117.
34. Marin, M. L. and Duran-Vila, N. (1988). *Plant Cell Tiss. Organ Cult.*, **14**, 51.
35. Withers, L. A. (1979). *Plant Physiol.*, **63**, 460.
36. Dereuddre, J., Blandin, S., and Hassan, N. (1991). *Cryo-Lett.*, **12**, 135.
37. Plessis, P., Leddet, C., and Dereuddre, J. (1991). *C. R. Acad. Sci. Paris*, S. III, **313**, 373.
38. Niino, T., Sakai, A., and Yakuwa, H. (1992). *Cryo-Lett.*, **13**, 51.
39. Uragami, A., Lucas, M. O., Ralambosa, I., Renard, M., and Dereuddre, J. (1993). *Cryo-Lett.*, **14**, 83.
40. Dereuddre, J., Blandin, S., and Hassan, N. (1991). *Cryo-Lett.*, **12**, 125.
41. Widholm, J. M. (1972). *Stain Technol.*, **47**, 189.
42. Steponkus, P. L. and Lamphear, F. O. (1967). *Plant Physiol.*, **42**, 1423.
43. Mannonen, L., Toivonen, L., and Kauppinen, V. (1990). *Plant Cell Rep.*, **9**, 173.
44. Benson, E. E., Lynch, P. T., and Jones, J. (1992). *Plant Sci.*, **85**, 107.
45. Benson, E. E. (1990). *Free radical damage in stored plant germplasm*, IBPGR, Rome, Italy.
46. Corbineau, F., Engelmann, F., and Come, D. (1990). *Plant Sci.*, **71**, 29.
47. Benson, E. E. and Withers, L. A. (1987). *Cryo-Lett.*, **8**, 35.
48. Benson, E. E. and Norhona-Dutra, A. A. (1988). *Cryo-Lett.*, **9**, 120.
49. Sowa, S. and Towill, L. E. (1991). *Plant Physiol.*, **95**, 610.
50. Harding, K. (1991). *Euphytica*, **55**, 141.
51. Ward, A. C. W., Benson, E. E., Blackhall, N. W., Cooper-Bland, S., Powell, W., Power, J. B., and Davey, M. R. (1993). *Cryo-Lett.*, **14**, 145.

Secondary products from cultured cells and organs: I. Molecular and cellular approaches

RICHARD J. ROBINS

1. Introduction

Over the past two decades, cell cultures have proved to be a valuable tool for the study of the biosynthesis of secondary products in plants. The term 'secondary products' covers a wide-range of chemically dissimilar compounds. These have in common a non-essential role for the survival of the individual plant cell, but, in contrast, where the *in vivo* role of these compounds has been ascertained, it is clear that they are often essential for the survival of the plant as a whole. Many of these compounds may not be synthesized during the entire lifetime of the plant. For example, flower pigments are only produced at a specific developmental stage, while a number of types of chemicals are rapidly synthesized, and as rapidly destroyed, in response to pathogenic invasion. Whatever their roles in the plant may be, however, mankind has found uses for numerous secondary products as pharmaceuticals or food ingredients.

This temporal and/or developmental segregation within the plant has often made the study of the biosynthesis of these compounds *in planta* difficult. Even for those secondary products that are synthesized more or less continuously, the rate of synthesis is frequently very low and the sites of synthesis can render studies of the biochemistry involved hard to perform.

Cell cultures have long been recognized as a means of avoiding many of these problems. It was argued that, by growing undifferentiated tissue *in vitro*, large amounts of biosynthetically active tissue could be generated under conditions in which both seasonality and tissue-specificity of production would be circumvented. By this means, it was considered feasible to grow large quantities of biomass for the biotechnological production of complex plant secondary products by fermentation. Regrettably, in practice this argument proved to be naive. While many plant species are amenable to being grown in culture, others are not. Furthermore, of those that could readily be grown under suitable conditions, many lacked the desired biosynthetic activity, or had only very low levels of production.

Although the biotechnological impact of plant cell cultures proved minimal, the culture systems developed in some species provided excellent material for biochemical studies (1). As anticipated, provided reasonable levels of product accumulation could be achieved, tissue culture systems have enabled major inroads to be made into the biosynthetic pathways of a number of important groups of metabolites. It is easy to manipulate culture conditions, to feed precursors, and to extract the tissue for enzymes and products. This has enabled aspects of the sequence of reactions, the genetic and biochemical regulation, and the morphological association of pathways to have been examined in tissue cultures.

Plant tissue cultures fall into a number of types:

(a) Callus cultures—these are mostly not suitable for studying secondary products.

(b) Dispersed cultures—these offer rapid growth but show genetic and biochemical heterogeneity, instability, and commonly lack expression of the desired secondary pathway. They can be 'induced' in some cases by media manipulation and/or elicitation. They are readily amenable to fermentation (see Chapter 9).

(c) Organ cultures—these may be excised roots, transformed roots, excised shoots, or transformed shoots. They show genetic and biochemical stability but are heterogeneous in having cells of many different types present. Usually, they exhibit the full biosynthetic capacity associated with the organ *in planta*. They are also amenable to fermentation (see Chapter 9).

Establishing any culture for the examination of secondary product formation will be affected by the species of interest. This chapter, therefore, can only give examples of systems in which such an objective has successfully been achieved. Since the first edition, organ cultures have been extensively developed and, in many cases, prove to be the tissue of choice. Examples of suspension cultures will be covered only briefly and the chapter largely will consider the application of organized cultures to metabolic studies.

2. Strategies for inducing secondary product formation in suspension cultures

2.1 Improvement of quinoline alkaloid production in *Cinchona ledgeriana* cultures

C. ledgeriana suspension cultures can be induced using sterile shoot cultures or plantlets germinated from surface sterilized seeds. Various combinations of 2,4-D, BAP, MS or B5, and sucrose should be tested (*Protocol 1*). Callus forms under a wide-range of conditions but shows the best sustained growth in our hands on B5 salts with 2,4-D at 2 mg/litre, BAP at 0.5 mg/litre, and sucrose at 20 g/litre. This same combination is suitable for developing a rather slow growing liquid suspension culture, which can be subcultured every 28 days with about 25-fold dilution. The culture is particulate, requiring cutting at each subculture. Over a 60-day period, 0.05 μg/g fresh mass (FM) alkaloid accumulates. To try and enhance production, the concentrations of first 2,4-D, and secondly BAP, can be varied, and the alkaloid content examined over four sequential 28-day growth periods.

It is important always to look at the long-term effect of alterations in the growth conditions. In some cases, no change may be seen in the first sub-culture but subsequently altered production can appear, indicating that the system is overloaded with the phytohormones of the first set of conditions. Alternatively, changes seen in the first subculture are sometimes not sustained, indicating that they may be a short-term response to an altered environment, not the establishment of a new steady-state set of biosynthetic conditions.

Yield is improved to about 20 μg/g FM by lowering the 2,4-D to 0.5 mg/litre and the BAP to 0.1 mg/litre. While slightly lower levels of these phytohormones give higher yields at the first subculture, growth is not sustained below 0.5 mg/litre 2,4-D. A further tenfold improvement can be achieved by testing different combinations of auxin and cytokinin. A final yield of 200 μg/g FM/28 days of growth can be attained on a medium containing 2 mg/litre indole acetic acid (IAA) and 0.1 mg/litre zeatin riboside (ZR). Up to 1000 μg/g FM accumulates in the first subculture.

The yield of this culture is adequate to examine enzymes involved in the biosynthesis of the quinoline alkaloids, notably tryptophan decarboxylase and quinidinone dehydrogenase. The culture is unstable, however, the alkaloid yield showing considerable variation with time. Also, growth in the optimal medium (IAA/ZR) is not sustained beyond five or eight subcultures. Furthermore, the maximum yield is only obtained during the first subculture into these conditions. Improved production and stability can be obtained by transforming *C. ledgeriana* with *Agrobacterium tumefaciens* or *A. rhizogenes* (2). Even these transformed cultures, however, prove to have rather slow growth rates and variable productive capacities.

Protocol 1. Maximizing quinoline alkaloid production in *C. ledgeriana* suspension cultures

A. *Establishment of culture*

1. Excise internode sections about 1 cm long from sterile shoot cultures (see *Protocol 8*).

2. Cut each section longitudinally and place cut surface down on semi-solid medium containing B5 or MS salt nutrients (see Chapter 1), 20 g/litre

Protocol 1. *Continued*

    sucrose, and various combinations of 2,4-D (range 0.05–5 mg/litre) and BAP (range 0.01–2 mg/litre). Incubate at 25 °C and illuminate (approx. 12 $\mu E \cdot m^{-2} \cdot s^{-1}$ fluorescent light) with 16 h light/day.

3. Remove the calli, which form after about three weeks, from these explants and subculture on to fresh medium.

4. Continue subculturing the healthy, growing calli at two to four weekly intervals.

5. To initiate the suspension culture, place 2–5 g FM callus into 50 ml liquid medium, of the same composition as the semi-solid medium on which it was growing, in a 250 ml Erlenmeyer flask.

6. Incubate on a gyratory shaker at about 90 r.p.m. at 25 °C and 16 h light/day.

7. Subculture every four weeks. Initially use a small dilution of three- to five-fold. Gradually increase this to 20- to 25-fold.

8. The *C. ledgeriana* cultures grow as large green-brown aggregates. To subculture, lift a number of these from the flask with a sterile stainless steel spoon and place on a 140 mm Petri dish. Slice the aggregates into pieces 1–2 mm thick, mix these together from several flasks, and transfer about 1 g FM to fresh medium.

B. *Improving productivity by varying conditions of growth and production*

1. Subculture tissue into a range of liquid media containing B5 salts, 20 g/litre sucrose, and 2,4-D (range 0.05–5 mg/litre) with BAP at 0.5 mg/litre.

2. Examine the alkaloid content in the first and three further subcultures into this medium.

3. Select the culture conditions showing the best sustained production of alkaloids.

4. Subculture tissue into a range of liquid media containing B5 salts, 20 g/litre sucrose, and BAP (range 0.05–5 mg/litre) with 2,4-D at 0.5 mg/litre.

5. Repeat steps **2** and **3**.

6. Examine the effect of using different cytokinins (typically BAP, kinetin, isopentenyladenine, ZR) at the optimal cytokinin concentration (0.1 mg/litre in this case), and auxins (typically 2,4-D, IAA, 4-chlorophenoxyacetic acid, 2-methyl-4-chlorophenoxyacetic acid, naphthaleneacetic acid (NAA), indolylbutyric acid (IBA), at the optimal auxin concentration (2.0 mg/litre for IAA, NAA; 0.5 mg/l for the other auxins used in this example).

7. Repeat steps **2** and **3**.

8. Grow cultures on best medium (in this case 2 mg/litre IAA and 0.1 mg/litre ZR).

C. Analytical procedure: extraction and quantitation of quinoline alkaloids

1. Harvest tissue aggregates and medium separately. Blot aggregates dry and store at $-20\,^\circ C$.

2. Grind tissue (0.5–5 g FM) with an Ultraturrax in 10 ml 0.2 M H_2SO_4, with 10 ml $CHCl_3$ (to remove pigments). Wash probe with 5 ml $CHCl_3$.

3. Stand for 90 min at room temperature.

4. Separate cell debris by filtration *in vacuo* through a glass fibre A paper (Whatman)/miracloth sandwich. Wash debris with 5 ml 0.2M H_2SO_4.

5. Centrifuge (1000 g; 5 min), separate layers, and discard $CHCl_3$ phase.

6. Make the aqueous phase alkaline with 1.5 ml 35% (v/v) NH_3 solution. Add 10 ml re-distilled diethyl ether or $CHCl_3$, flush with nitrogen gas, stopper, and leave overnight at $4\,^\circ C$.

7. Separate the phases and take the organic phase to dryness.

8. Take up the residue in 0.2 ml HPLC running mixture (see step **10**).

9. Treat media samples similarly except add 2.5 ml 0.2 M H_2SO_4 to each 10 ml of medium.

10. Inject 20 µl of extract on to a µBondapak-C18 column (Waters) and develop isocratically with acetonitrile/water/acetic acid/tetrahydrofuran (50:450:5:2) at 1.4 ml/min, detection at 240 nm. Quinoline alkaloids elute in the order cinchonine (8.3 min), cinchonidine (9.4 min), dihydrocinchonine (11.0 min), quinidine (11.8 min), quinine (14.2 min), dihydroquinidine (16.1 min), dihydroquinine (19.9 min).

2.2 Elicitation of acridone epoxide alkaloids in *Ruta graveolens* suspension cultures

In a number of suspension cultures, the level of desired secondary product accumulated can be dramatically increased by elicitation. The response may be induced by UV light, fungal or bacterial culture filtrates, polymers like chitosan, heavy metals, or even dilution into fresh medium. Usually, though not always, this process is most effective for compounds that are naturally phytoalexins, i.e. that are *de novo* synthesized defence compounds *in planta*. In these cases, induction is rapid and may be transient, a peak of product accumulation appearing within 24 to 48 hours of elicitation. In other cases, where the response may be due to less specific inducibility of the pathway, a slower accumulation occurs, possibly reaching a maximum after five to ten days.

Suspension cultures of *R. graveolens* are induced by micro-organisms, such as *Rhodotorula rubra*, to accumulate acridone epoxide alkaloids. This event is associated with a rapid and transient induction of the expression of S-adenosyl methionine:anthranilic acid *N*-methyltransferase (NMT), the first enzyme in the pathway. *Protocol 2* illustrates an example of this induction. It also shows how to indicate simply that the induction is due to a diffusible molecule by immobilizing the plant and fungal cultures.

Protocol 2. Elicitation of *R. graveolens* suspension cultures

1. Establish a callus essentially as described in *Protocol 1*, steps 1 to 3.

2. Propagate callus photomixotrophically on modified Eriksson medium, at 25 °C, illuminated at 25–65 μE·m^{-2}·s^{-1} fluorescent light. Subculture green callus every three to six weeks.

3. Establish suspension cultures by subculturing light grown, green callus into MS medium containing 0.1 mg/litre 2,4-D and 0.25 mg/litre kinetin. Grow for several generations at 25 °C and 16 h light/day.

4. Grow culture of *Rh. rubra* in MS medium for seven days to a density of about 40 mg FM/ml.

5. Mix 40 ml sterile sodium alginate (35 g/litre) with 40 ml *Rh. rubra* suspension containing 1.6 g of the yeast.

6. Pour dropwise into a beaker of gently stirred 70 mM CaCl$_2$. Beads 2–3 mm in diameter should form.

7. Leave to cure for 2 h and wash three times with fresh medium.

8. Mix 40 ml sodium alginate (35 g/litre) with 40 ml *R. graveolens* suspension culture containing about 20 g FM.

9. Repeat steps **4** and **5**. Beads 3–5 mm in diameter should form.

10. For control treatments, prepare beads as in steps **3** to **5** but lacking yeast cells.

11. Place *R. graveolens* beads into 300 ml Erlenmeyer flasks containing MS medium and add 1 g *R. rubra* beads.

12. Co-cultivate for various periods up to 120 h, taking samples.

13. Analyse samples for acridone oxides and NMT activity (see Eilert in ref. 1).

Suspension cultures of *R. graveolens* grown in the dark accumulate higher levels of acridone oxides than those cultured in the light. Also, these cultures produce furoquinoline alkaloids and furanocoumarins. Both these groups of compounds are elicited by *Rh. rubra*.

3. Strategies for establishing excised root-organ cultures competent in secondary product formation

Although the basic technique of excised root culture was established by 1964, little attention had been paid to the potential of this approach for secondary metabolites until the early 1980s. Furthermore, the range of roots that could be readily grown in this way was limited. Methods by which the range can be enhanced by using genetic manipulation are discussed in Section 4.

3.1 Establishment of tropane alkaloid producing cultures of *Hyoscyamus niger* and *H. albus*

Attempts to produce tropane alkaloids in callus or suspension cultures have met with little success; only in a few species have measurable levels been obtained. In contrast, when excised roots of *Hyoscyamus* species are established in culture, good levels of production result.

Cultures produce alkaloids continuously during the culture period, with a maximum content of about 1% dry mass (DM) in *H. albus* and 0.1% DM in *H. niger*. The effect of the phytohormones IAA, NAA, IBA, and 2,4-D over the range 0.1 to 100 μM has been tested. By supplementing the medium with IBA at 10 μM, the growth rate of both cultures is increased by about 50%. This is apparently due to a stimulation in the formation of lateral root primordia. Increasing the growth rate, however, causes a decrease in alkaloid accumulation. All four phytohormones cause a decrease in the total alkaloid content of both species, although some effects on alkaloid composition are also seen.

Protocol 3. Establishment and analysis of a *Hyoscyamus* root culture

A. *Establishment of cultures*

1. Germinate sterile seeds to obtain sterile seedlings.

2. Excise a small part of the cotyledon and part of a root about 5 mm long from each seedling and place in Linsmaier and Skoog medium with 30 g/litre sucrose at 25 °C in the dark on a reciprocal shaker at 60 strokes/min.

3. Leave to grow and, once established, subculture about 1 g FM roots at two-weekly intervals into fresh medium. Cultures increase about four- to sixfold in mass over this period.

B. *Analytical procedure*

1. Harvest, blot, and freeze-dry root tissue.

2. Powder and soak overnight in EtOH/28% NH_3 solution (19:1) mixture. Centrifuge (5 min, 1500 g) and twice repeat the extraction of the pellet.

3. Pool the alkaline alcoholic extracts and take to dryness at 40 °C.

4. Dissolve the residue in 2 ml 0.1 M HCl and filter. Make 1 ml alkaline with 0.2 ml 1 M Na_2CO_3/$NaHCO_3$ and apply to 1 g Extrelute® column (Merck). After 5 min elute with 6 ml $CHCl_3$.

Protocol 3. *Continued*

5. Evaporate CHCl₃ to dryness and dissolve the residue in 1,4-dioxane and *N,O*-bis(trimethylsilyl)acetamide (4:1 v/v), containing tricosan at 0.5 mg/ml as internal standard.

6. Apply 1–5 µl to an OV-101 (25 m × 0.3 mm) capillary gas chromatography (GC) column at 250 °C, with a He flow of 1 ml/min, and a split ratio of 50:1. Record the separation by flame ionization detector (FID). Under these conditions the alkaloids elute in the order hyoscyamine (4.51 min), littorine (4.61 min), scopolamine (5.54 min), and 6β-hydroxyhyoscyamine (5.89 min).

3.2 Establishment of pyrrolizidine alkaloid producing cultures of *Senecio* species

A number of species of *Senecio* (*S. vulgaris*, *S. vernalis*, *S. erucifolius*, *S. squalidus*) have been established as root cultures. Like *Hyoscyamus*, these are readily established and provide rapidly growing cultures. These cultures accumulate a number of pyrrolizidine alkaloids, primarily as their *N*-oxides. They have proved valuable for studies of the incorporation of putative precursors such as ornithine, arginine, putrescine, isoleucine, and spermidine into these bases.

Protocol 4. Establishment and analysis of *Senecio* root cultures

A. *Establishment of cultures*

1. To obtain sterile seedlings, germinate sterile seeds on MS medium containing 40 g/litre sucrose and no phytohormones.

2. Excise several roots of at least 1–2 cm length and incubate in 5 ml MS medium containing 40 g/litre sucrose.

3. After two to three passages in this small volume, place the cultures in 50 ml MS medium in 200 ml Erlenmeyer flasks on a gyratory shaker at 120 r.p.m. at 25 °C.

4. Transfer the roots (1.5–2 g FM) to fresh medium every 14 days.

B. *Analytical procedure 1 (total alkaloids)*[a]

1. Homogenize (Ultraturrax) plant material (6–10 g FM) in 20 ml 0.05 M H₂SO₄ and leave to stand for 30 min.

2. Centrifuge, divide the supernatant in half and treat as follows:
   (a) To obtain the tertiary pyrrolizidine alkaloids make alkaline with NH₃ solution and apply to an Extrelute column at 1.4 ml/g resin. Elute with CH₂Cl₂ (6 ml/g Extrelute), evaporate to dryness, and dissolve the residue in MeOH.

(b) To obtain the quaternary pyrrolizidine alkaloid *N*-oxides adjust to 0.25 M H_2SO_4, mix with an excess of zinc dust, and stir for 5 h at room temperature. Make alkaline with NH_3 solution and proceed as in (a).

3. Apply 1–2 μl to a DB-1 (25 m × 0.3 mm) capillary GC column at 120 °C, injector at 250 °C, with a He inlet pressure of 0.7 bar, and a split ratio of 50 : 1. Develop the chromatogram from 120 to 290 °C, at 8 °C/min. Record the separation by FID or, for more sensitivity, by phosphorus/nitrogen detector (PND).

C. Analytical procedure 2 (N-oxides)[a]

1. Extract (Ultraturrax) freeze-dried roots with MeOH (30 ml/0.5 g DM) for 3 min. Centrifuge. Take the supernatant to dryness and dissolve the residue in MeOH.

2. Inject 20 μl on to a μBondapak C-18 column (300 mm × 3.9 mm) and develop isocratically with 15 mM potassium phosphate buffer pH 7.5/ MeOH (2 : 1), at a flow rate of 1.5 ml/min, and a detector at 209 nm. The pyrrolizidine alkaloid *N*-oxides elute in the order retrorcine *N*-oxide (3.9 min), seneciphylline *N*-oxide (6.1 min), senecionine *N*-oxide and integerimine *N*-oxide (8.0 min). The pyrrolizidine alkaloids can also be separated by using the buffer at a 1:1.5 ratio.

[a] The analytical procedure described here is specially developed for the extraction and quantification of alkaloid extracts where there is a significant amount of alkaloid present as the *N*-oxide.

4. Strategies for establishing transformed root-organ cultures competent in secondary product formation

In *Agrobacterium*-mediated transformation, a small segment of DNA, the transfer (T-)DNA, is copied from a bacterial plasmid and transferred to the plant (3). In the plant cell it becomes integrated into the genome. The T-DNA carries on it a number of open reading frames with prokaryotic promoters that express in eukaryotic organisms to influence plant development. In the case of *A. rhizogenes*, three genetic loci, *rolA*, *rolB*, and *rolC*, have been identified that are responsible for the establishment and maintenance of the rooty phenotype. Transformation provides a method by which roots of a wider range of species can be generated and cultured in axenic conditions. It has the advantages that:

- cultures do not need added phytohormones
- cultures are frequently fast growing (though not always)
- genetic material can easily be introduced (see Section 8)

4.1 Establishment of pyridine alkaloid producing cultures of *Nicotiana rustica*

The genus *Nicotiana* is widely used in plant biology. In addition, the pyridine alkaloids, notably nicotine, are synthesized and accumulated in the roots. Cultures of *N. rustica*, generated by transformation with *A. rhizogenes*, have proved valuable for:

- studying the flexibility of the roots in terms of their being able to metabolize exogenous substrates
- examining the effect on nicotine production of genetically engineering the pathway (see Section 8.1)
- investigating the relationship between alkaloid accumulation and morphological integrity (see Section 7.2)

Protocol 5. The generation of transformed roots of *N. rustica*

A. *Establishment of cultures*

1. Surface sterilize seeds with 20% (v/v) commercial bleach for 30 min. Wash six times with 100 ml sterile tap-water.
2. Germinate sterile seeds on MS medium containing 30 g/litre sucrose, 8 g/litre agar, and no phytohormones, and grow under fluorescent lighting, 16 h day, 25 °C.
3. Prepare a 48 h suspension of *A. rhizogenes* LBA9402 in TYM medium at 25–26 °C. (*Note*: temperatures above 28 °C tend to cure the bacterium of the plasmid.)
4. Wound the stem of six- to eight-day-old seedlings with a narrow hypodermic needle containing the *A. rhizogenes* suspension.
5. When emergent roots are 5–10 mm long (7 to 14 days), excise the stem segment bearing roots, and place in 8 ml B5 medium containing 30 g/litre sucrose, and 0.5 mg/ml ampicillin sulfate.
6. Passage rapidly growing roots (approx. 0.5 g) into 50 ml of the same medium in 250 ml Erlenmeyer flasks.
7. Continue subculturing every two weeks. Passage two to four lengths of root (3–4 cm long and bearing numerous side-branches) into fresh medium.
8. After about eight subcultures it should be possible to omit ampicillin from the medium. If problems are encountered, proceed as in *Protocol 6*.

B. *Analytical procedure*

1. Homogenize (Ultraturrax) root material (1–4 g FM) in 15 ml 2% (v/v) NH_3 solution and 15 ml $CHCl_3$ and leave to stand for 90 min.

---

2. Vacuum filter through a glass microfibre A (Whatman)/miracloth sandwich.

3. Separate the phases. It may prove necessary to centrifuge the emulsion at 700 g for 5 min to achieve separation. Retain the $CHCl_3$.

4. Treat 10 ml samples of culture medium with 1 ml 35% (v/v) NH_3 solution and 15 ml $CHCl_3$ and handle as in steps **1** to **3**.

5. Treat the $CHCl_3$ phase with 0.3 ml/ml of 0.2 M H_2SO_4. It may prove necessary to centrifuge as in step **3**. Retain the aqueous phase.

6. Neutralize an aliquot of the aqueous phase with 35% (v/v) NH_3 solution and use directly for HPLC analysis.

7. Inject 10–50 µl on to a µBondapak-C18 column and elute isocratically with acetonitrile/water/acetic acid/tetrahydrofuran (12:430:3:1) pH 4.0 at 1.0 ml/min, detection at 260 nm. Tobacco alkaloids elute in the order nornicotine (7.7 min), anatabine (8.9 min), nicotine (9.6 min), anabasine (11.8 min).

Protocol 6. Decontamination of cultures using an antibiotic mixture

1. Prepare a solution of 25 mg/litre ampicillin sulfate, 25 mg/litre cephalosporin C, and 25 mg/litre cephotaxime in water.[a] Filter sterilize, aliquot in 1 ml portions, and store frozen.

2. Remove contaminated culture from flask, draining off as much liquid as possible.

3. Rinse for about 10 sec in a solution of 25 mg/litre ampicillin sulfate.

4. Wash culture in seven sequential 50 ml aliquots of sterile distilled water.

5. Place into 50 ml fresh medium containing 0.25 ml of the antibiotic mix.

6. Leave for one to five days in this medium, examining daily.

7. If there is any sign of toxicity to the culture, remove immediately to fresh medium containing 0.5 ml per 50 ml of ampicillin sulfate (25 mg/ml stock).

8. Subculture to medium lacking antibiotics and examine for contamination.

[a] This mixture of antibiotics is harsh and should only be used in emergencies. Cephotaxime in particular can be quite toxic to plant cells. Any bacterial contaminants which can survive this treatment should be destroyed by autoclaving the sealed flask.

4.2 Establishment of tropane alkaloid producing cultures of *Datura stramonium*

Root cultures of *D. stramonium* and a range of other *Datura* species have proved invaluable in studies of the biosynthesis of tropane alkaloids (see Section 6). These cultures are easy to establish and fast growing. They provide a good system for precursor and inhibitor feeding and are an excellent tissue from which to extract enzymes.

Protocol 7. The production of transformed roots of *D. stramonium*

A. *Establishment of cultures*

1. Surface sterilize seeds with 10% commercial bleach for 30 min. Wash six times with 100 ml sterile deionized water.

2. Germinate sterile seeds on B5 medium containing 10 g/litre agar, no sucrose, and no phytohormones.

3. Prepare a 48 h suspension of *A. rhizogenes* LBA9402 in TYM medium at 25–26 °C.

4. Wound the stem of two- to three-week-old seedlings[a] with a narrow hypodermic needle containing the *A. rhizogenes* suspension. Initiate and maintain root cultures as in *Protocol 5A*, steps 5–8.

B. *Analytical procedure*

1. Homogenize (Ultraturrax) root material in 0.2 M H_2SO_4 (5 ml/g FM) and leave to stand for 90 min.

2. Vacuum filter through a glass microfibre A/miracloth sandwich.

3. Extract the filtrate with an equal volume of $CHCl_3$. It may prove necessary to centrifuge the emulsion at 700 g for 5 min to separate the phases. Retain the aqueous phase.

4. Make the aqueous phase alkaline with 35% (v/v) NH_3 solution and extract with an equal volume of $CHCl_3$. It may prove necessary to centrifuge the emulsion at 700 g for 5 min to separate the phases.

5. Evaporate to dryness and dissolve the residue in 1 ml MeOH.

6. Dilute 50 µl of the methanolic solution into 1 ml ethyl acetate. Inject 1–5 µl on to a DB-17 (25 m × 0.32 mm, film thickness 0.23 µm) capillary GC column at 65 °C, fitted to a gas chromatograph with a cold on-column injector (e.g. Carlo Erba Mega MR5160), with a He flow of 1.5 ml/min.[b] Develop the chromatogram from 65 to 120 °C at 4 °C/min, from 120 to 280 °C at 8 °C/min, and from 280 to 300 °C at 12 °C/min, with 4 min at 300 °C. Record the separation by PND. Under these conditions the alkaloids elute in the order tropine (13.8 min), tropinone (14.0 min), pseudotropine (14.5 min), acetyltropine (17.0 min), cuscohygrine (23.7 min), tiglyltropine (25.0 min), littorine (32.0 min), hyoscyamine (33.5 min).

---

[a] Seedlings of different species may require longer to reach a suitable size. Alternatively, leaf or stem tissue can be used. Care must be taken not to apply to the tissue more bacterial suspension than can be held in the surface scratch, otherwise the bacterial growth can overrun the emergent roots and make it very hard to obtain axenic cultures. If difficulties are encountered then it may

prove possible to clear the culture of bacteria with a strong antibiotic mix (*Protocol 6*). A useful approach is to use a strain of *A. rhizogenes* carrying a binary vector with the coding sequence of an antibiotic marker gene. It is then possible to select for the transformed roots by their ability to grow in the presence of the antibiotic. For example, *A. rhizogenes* AR1601 confers resistance to kanamycin to the transformed tissue. This approach is particularly valuable with species prone to form adventitious roots in response to wounding. The time taken for roots to emerge from the wound site can vary considerably, from a few days to several months. Roots are unlikely to be produced once the parent tissue is fully necrotic but can emerge at a late stage of tissue decay.

[b] The use of a cold on-column injector is desirable with tropane alkaloids. At the high injector temperatures (250 °C) required to volatilize hyoscyamine and the other high molecular weight esters (e.g. scopolamine), some degradation occurs and this can lead to inaccurate quantitation as the extent of degradation varies with the composition and concentration of the sample being applied.

5. Strategies for establishing shoot-organ cultures competent in secondary product formation

As with root cultures, some shoot cultures (e.g. *N. tabacum*) can be established extremely easily, simply by removing the top from sterile seedlings and placing in B5 or MS medium with 10 g/litre sucrose. In other cases, however, a more complex procedure and medium are required.

5.1 Induction and propagation of shooty cultures of *C. ledgeriana*

A number of shoot cultures of *C. ledgeriana* have been described. The method given here has been found to establish cultures that can be maintained for a long period of time. It is suitable for the establishment of shooty cultures of a number of woody species. Shoot cultures of *C. ledgeriana* maintained under these conditions are competent in quinoline alkaloid production.

Protocol 8. Formation and analysis of shooty cultures of *C. ledgeriana*

1. Germinate seedlings in greenhouse conditions.

2. Excise the terminal 15 mm portion from 30-week-old seedlings.

3. Dip into 0.1% (v/v) Nonidet and sterilize by submerging in sodium hypochlorite (0.15% available chlorine) for 1 min, and rinse three times in sterile water.

4. Implant singly in 25 ml MS medium containing 10 g/litre agar in a 50 ml Erlenmeyer flask, and incubate in continuous light at 22 °C for 20 h.

5. Remove explant from medium, dip in 0.1% (v/v) Nonidet, and sterilize again by submerging in sodium hypochlorite (0.45% available chlorine) for 40 min.

Protocol 8. *Continued*

6. Excise the terminal 6 mm of the explant and implant into MS medium supplemented with 10 g/litre agar, 1 mg/litre gibberellin, 1 mg/litre IBA, 120 mg/litre phloroglucinol, and 10 mg/litre powdered activated charcoal.

7. Culture under red-enhanced 'Growlux' illumination at 22–26 °C with a 16 h photoperiod. Subculture every six to eight weeks by excising the terminal 10–15 mm and implanting in fresh medium.

8. Assess quinoline alkaloid production as described in *Protocol 1*.

5.2 Establishment of transformed shoot-organ cultures of *Mentha citrata* and *M. piperita*

An alternative approach to forming sustainable cultures of shooty tissue is to transform with *A. tumefaciens*. This soil pathogen, like *A. rhizogenes*, also carries a plasmid that bears genes which, when transferred into the plant, cause altered phenotype (3). In this case, the normal phenotype is a callus, the so-called 'crown gall'. Mutants of this bacterium have been made by transposon mutagenesis of the phytohormone genes of the T-DNA. On *N. tabacum*, these cause shooty or rooty cultures to appear, depending on whether the insertion is in the auxin (*tms*) or cytokinin (*ipt*) loci respectively. In theory, therefore, these mutants should be suitable to produce shooty cultures. In practice, this phenotype is frequently not shown by the auxin mutants on a range of other species: indeed, some rooty cultures are formed. Surveying a range of hosts and pathogens, Rhodes *et al.* (2) demonstrated that certain wild-type strains of *A. tumefaciens* would induce shoots under appropriate conditions. In particular, the nopaline strains T37, C58, and N273 induced shoots on *M. citrata*, *M. piperita*, or both. Since shooty phenotypes tend to be generated by enhanced cytokinin levels, constructs have also been made using the disarmed strain LBA4404 carrying the *ipt* gene expressed under the powerful cauliflower mosaic virus (CaMV) 35S protein promoter. This induces high cytokinin production in transformed tissue and a shooty phenotype.

Protocol 9. The formation of transformed shoots of *M. citrata*

A. Establishment of the culture

1. Surface sterilize stem sections of greenhouse grown plants with 10% (v/v) commercial bleach for 10 min. Wash six times with sterile deionized water (100 ml).

2. Prepare a 48 h suspension of the appropriate strain of *A. tumefaciens* in TYM medium at 26 °C.

3. Wound the apical end of the stem with a narrow hypodermic needle containing the *A. tumefaciens* suspension. Take care not to over-infect and to keep the bacterial suspension well away from the basal end of the tissue. If too many bacteria are present they overgrow on the surface of the medium and tend to make it extremely difficult to establish an axenic culture.

4. Place the basal end of the inoculated stem into 50 ml 10 g/litre agar medium containing MS salts and 30 g/litre sucrose in a 200 ml jar.

5. Maintain under illumination, at approximately 25 $\mu E \cdot m^{-2} \cdot s^{-1}$ fluorescent light, 16 h day, 25 °C.

6. Prepare a conditioning medium by growing a suspension culture of *N. tabacum* in MS medium with 30 g/litre sucrose and 0.5 mg/litre 2,4-D for seven days. Allow the cells to settle out and harvest the medium. Store sterile and frozen until required.

7. After three to four weeks, excise the galls that appear at the points of infection and transfer to 10 g/litre agar medium containing MS salts, 30 g/litre sucrose, 1 mg/ml carbenicillin, and 10% (v/v) conditioning medium in Petri dishes.

8. After three to four days, transfer to fresh medium.

9. Continue to transfer to fresh medium every 21 days. Shoots should emerge from the galls after about four weeks in culture.

10. Excise individual shoot tips and culture separately on the MS/conditioning medium.

11. Maintain by subculturing approximately 1 g of shooty teratoma on to fresh MS medium approximately every three weeks.[a]

12. Establish liquid cultures by subculturing approximately 1 g of shooty teratoma into 50 ml B5 salts with 30 g/litre sucrose and growing on a gyratory shaker at 25 °C, 100 r.p.m., 25 $\mu E \cdot m^{-2} \cdot s^{-1}$ fluorescent light.

B. *Analytical procedure*

1. Harvest tissue (leaf and stem separately), freeze in liquid N_2, and store at −70 °C until required.

2. Weigh sample (approx. 1 g) into a pre-cooled (dry ice) centrifuge tube.

3. Add 1 ml cold (dry ice) hexane (HPLC grade), and homogenize.

4. Leave 15 min to extract and centrifuge at 3000 g for 10 min at 4 °C.

5. Repeat steps 3 and 4 a further three to six times.

6. Pool the supernatants in a cold (dry ice) vial and store at −20 °C until required.

7. Inject 1–2 μl on to a CP52 aluminium-clad capillary GC column (Chrompak: WCOT fused silica 25 m × 0.2 mm, film thickness 0.22 μm) using an on-column injector at 70 °C. Develop with He gas at 0.65 bar (10 ml/min)

Protocol 9. *Continued*

from 70 °C to 195 °C at 5 °C/min with 5 min at 195 °C. Detect the products with a FID. Terpenes of *M. citrata* elute in the order linalool (6.2 min), linalyl acetate (6.5 min). Terpenes of *M. piperita* elute in the order 1-menthone (4.9 min), menthofuran (5.0 min), 1-menthyl acetate (6.6 min), 1-menthol (7.9 min).

---

[a] Obtaining shooty teratomas completely free of traces of contaminating *A. tumefaciens* can be extremely difficult. Even after numerous passages on agar, bacterial outbreaks may occur. Treating liquid cultures with 3 mg/litre ampicillin sulfate appears to remove all traces of bacteria from the teratomas. Alternatively, treat as in *Protocol 6*. To confirm the absence of bacteria, use polymerase chain reaction (PCR) and amplify with primers for the *vir* gene region of the plasmid. This should be negative. Simultaneously, amplify the *rol* gene region of the T-DNA. This should be positive.

6. Use of cultures for studies of the biochemistry of secondary product formation

Cultures offer the opportunity to expose tissue efficiently and evenly to exogenous effectors or substrates. Tracer experiments with either radio- (e.g. ^{3}H or ^{14}C) or heavy isotopes (e.g. ^{2}H, ^{13}C, or ^{15}N) can often lead to far higher levels of specific incorporation in the compounds under study than is achieved using whole plants. Provided the tracer is rapidly absorbed, the kinetics of its incorporation and turnover can also be determined. Such studies can even be performed using *in vivo* nuclear magnetic resonance spectroscopy (NMR). Substrates, either normal components of the pathway or analogues of these compounds, can easily be introduced into the system and their metabolism, or effect on the metabolism of other compounds, be examined. By using a tissue that is axenic, possible interference by contaminating micro-organisms is avoided. Numerous examples of these types of studies exist, using callus, suspension, or organized culture tissue. An example will be given of the use of *Datura* roots to study flux in the alkaloid pathway.

Additionally, cultures provide a good source of enzymes. In part, their value is enhanced by having relatively thin cell walls, making the disintegration of the tissue easier than often is the case for plant organs. Furthermore, provided the culture is of high synthetic capacity, it may contain a greater proportion of actively synthesizing cells than would normally be the case in the plant.

6.1 Feeding of precursors

6.1.1 Flux analysis with known intermediates

The kinetics of alkaloid formation in transformed roots of *Datura* (see Section 4.2) has been established by examining the alkaloid and polyamine profiles in growing root cultures (see *Figure 1*). From these data, the sizes of the pools of

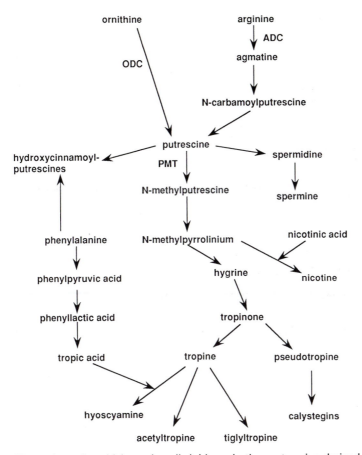

Figure 1. The pathway by which tropine alkaloids and other putrescine-derived metabolites are formed.

intermediates can be determined. In order to probe the regulation of the pathway, in particular which enzymic steps might limit the flux of metabolites into the end products (hyoscyamine, tropine esters), a series of experiments was performed in which different intermediates and precursors were fed and their incorporation into other pathway components measured (see *Protocol 10*). It is also valuable to see the effect that common precursors have on the pools of those compounds early in the pathway that are also metabolized to other groups of products (see Section 6.1.3). With the *Datura* roots, it was found that none of the intermediates fed had a stimulatory effect on the level of hyoscyamine present. Most of the compounds fed, however, did enhance the pool sizes of intermediates of the pathway, down as far as tropine, indicating that the esterification of tropine with tropic acid is a potentially flux-limiting step.

Protocol 10. The feeding of precursors and pathway intermediates

1. Subculture root cultures (see *Protocol 7*).

2. Prepare concentrated stock solutions of precursors at 55 mM or 110 mM in distilled water. Neutralize and make to volume. Filter sterilize.

3. Add required volumes, under sterile conditions, to each flask at required time intervals (determine experimentally).

4. Harvest tissue at required time intervals (determine experimentally) and extract alkaloids (see *Protocol 8*).

6.1.2 Studies of the pathway using specific enzyme inhibitors

Inhibitors highly specific to particular enzymes can be valuable tools to the biochemist who wishes to test the importance of that particular activity *in vivo*. Such inhibitors have been developed for both ornithine decarboxylase (ODC) and arginine decarboxylase (ADC); these are α-difluoromethyl-DL-ornithine (DFMO) and α-difluoromethyl-DL-arginine (DFMA) respectively. These compounds are catalytically activated inhibitors—so called 'suicide' substrates—and irreversibly inactivate the enzyme by covalently labelling the active site. When fed to *D. stramonium* roots, it has been found that DFMA, but not DFMO, inhibits the formation of hyoscyamine. From these data it is possible to infer that ADC is more important for tropane alkaloid formation than is ODC. In contrast, the synthesis of the polyamines spermine and spermidine are maintained in the presence of either inhibitor, suggesting that the biosynthesis of these primary metabolites is more important to the survival of the cells.

6.1.3 Use of NMR to establish potential intermediates

With the tropane alkaloids, because they are analysed by GC (see *Protocol 7*), it is possible to perform very sensitive incorporation experiments with heavy-labelled stable isotopes. The incorporation into all the intermediates can be followed by GC–mass spectrometry (MS) and the distribution of label within the major products determined, following their isolation, by NMR. This approach has been taken with *D. stramonium* roots to which [1,2-^{13}C]acetate or [2',3'-^{13}C]hygrine have been fed to probe the tropine pathway, or [1,3-^{13}C]phenyllactic acid was fed, to probe the tropic acid pathway (see *Protocol 11*). Competition experiments can also be performed, in which the potential precursor is fed with compounds that are putative later intermediates of the pathway. These should diminish incorporation.

Protocol 11. The feeding of labelled [1,3-^{13}C]phenyllactic acid

1. Subculture root cultures (see *Protocol 7*). Leave three or four days to allow growth to be established and to confirm sterility.

2. Prepare a stock solution of 110 mM [1,3-^{13}C]phenyllactic acid in distilled water.[a] Neutralize and make to volume. Filter sterilize.

3. Under sterile conditions, add 1 ml to each flask, giving a final concentration of 2 mM.

4. Repeat feeding twice more at two or three day intervals.

5. Harvest tissue at 10 to 12 days post-subculture and extract alkaloids (see *Protocol 7*).

6. To analyse by GC–MS, use a system such as a DB1 or DB17 column (30 m × 0.32 mm, film thickness 0.25 μm) fitted to a Hewlett Packard 5890 GC linked to a TRIO 1S mass spectrometer.

[a] Potential precursors are usually prepared as a 20- to 50-fold concentrated sterile stock solution in water, so that only 1–5 ml/flask are added. For less water soluble compounds, use a non-aqueous solvent such as methanol, ethanol, or dimethylsulfoxide. The final concentration should not exceed 5–10% (v/v) at which level these solvents are normally tolerated by the cultures.

6.2 Extraction and assay of enzymes

Root cultures provide good quality material for the preparation of enzymes. The tissue has the advantages that:

(a) It is not all highly vacuolated, giving a good protein : fresh weight ratio.

(b) It is not highly lignified, making it relatively easy to obtain good disintegration.

(c) It can be obtained in large quantities by fermentation, making it useful for large-scale enzyme preparations.

(d) It is not strongly pigmented, although it can contain high levels of phenolic compounds.

While there are many different ways to extract enzymes from plant tissue, *Protocol 12* gives a generally applicable set of conditions that we have developed and which appear to give good reproducibility and recovery. The buffers and specialist additives required will vary with the enzyme being prepared. However, it is advisable to:

(a) Use a neutral or slightly alkaline buffer with an ionic strength of at least 100 mM and preferably 200 mM.

(b) Include a chelating agent (usually EDTA) at 5–20 mM to remove heavy metals.

(c) Include a reducing agent (e.g. dithiothreitol, mercaptoethanol) at 3–20 mM to maintain thiol groups in the reduced state and to minimize polyphenol oxidase activity.

(d) Add an adsorbent (e.g. Polyclar AT, XAD-7) to remove phenolics from the extract (otherwise these tend to bind to and inactivate the enzymes present).

(e) Carry out all stages of the extraction in a 4 °C cold room. Otherwise, keep all buffers, apparatus, etc. cold with ice.

Protocol 12. The preparation of enzymes from transformed roots

1. Subculture roots.

2. At the desired age, harvest the tissue, blotting thoroughly on a paper towel.

3. Weigh the tissue and make into packets containing approximately 20 g (small scale preparation) or 60 g (large scale preparation).

4. Freeze rapidly in liquid nitrogen.

5. Store in ultra-cold conditions, preferably −70 or −40 °C. Tissue seems stable for at least one year under these conditions.

6. Remove packets from freezer into liquid nitrogen.

7. Prepare a beaker of buffer, containing about 3 ml buffer/g FM of tissue to be extracted, stirring gently on a magnetic stirrer.

8. Weigh out an aliquot of Polyclar AT at approximately 1 g Polyclar AT/5 g FM roots in the packet to be handled next.

9. Remove a packet of frozen roots one at a time from the liquid nitrogen, crush vigorously on a hard, cold surface. We use a marble pastry board and rolling pin.

10. Tip the crushed root material into the cold bowl of a coffee grinder, adding the Polyclar AT, mix briefly, and grind—preferably shaking the grinder—for not more than 5–10 sec. Do not allow to thaw.

11. Tip the ground powder into the moving buffer and stir gently with a glass rod to mix.

12. Repeat steps **8** to **11** until all the tissue is disintegrated.

13. Leave about 30 min, stirring gently.

14. Press out the debris from the suspension, in aliquots, either by squeezing through linen or miracloth, or with a mechanical fruit press.

15. Centrifuge the liquid at 15 000 g, 15 min, at 4 °C.

16. Decant the supernatant.

17. To determine the activity in the crude extract, desalt 2.5 ml on a PD-10 column (Pharmacia) or equivalent.

As an example, the time courses of ornithine and arginine decarboxylase activities and the effect of precursors and inhibitors on enzyme levels in transformed roots of *D. stramonium* have been determined. Using a protocol like *Protocol 12*, the activity of these enzymes in tissue of different ages can be examined. Both activities, and that of several other enzymes of tropane alkaloid formation, are found to have a maximal specific activity in roots harvested at ten days old. Therefore, for enzyme preparations, roots of this age are used. It should be noted that the total maximal extractable activity per flask (or fermenter) may be greater in slightly older tissue, depending on the growth characteristics of the culture and the acuteness of the maximum observed. However, it is usually advisable to purify enzymes using the tissue having the highest specific activity of the enzyme.

In studying the regulation of tropane alkaloid formation, it is desirable to examine the effect of high levels of intermediates of the pathway and inhibitors, such as DFMO and DFMA, on the levels of these activities. Grow roots in the presence of 1–10 mM effector and harvest at ten days. Following extraction essentially as above, pass the supernatant from *Protocol 12*, step 17, down a second PD-10 column. This is to ensure that all the added effectors are removed from the extract: one passage proves to be insufficient for cultures fed at high levels of effector and interference is caused in the assay of the enzyme activity. From this information, it is possible to conclude, for example, that putrescine and agmatine have a much greater effect on the level of arginine decarboxylase activity than on the level of ornithine decarboxylase activity.

Unexpectedly, feeding DFMA or, to a lesser extent, DFMO, is found to decrease the level of S-adenosyl methionine : putrescine *N*-methyltransferase (PMT), an enzyme for which these compounds are not inhibitors. Following up this observation, it is apparent that the level of agmatine being synthesized might be having a feed-forward effect on the level of expression of PMT. The observed effect of DFMA is overcome by adding exogenous agmatine to the cultures. Thus, inhibitor studies are sometimes able to give an indication of the regulation of the pathway in addition to the likely role within it of the target enzyme.

7. Influence of morphology on secondary product formation

As indicated in the Introduction, the morphology of a culture can strongly influence the ability of the tissue to biosynthesize secondary products. In this section, examples will be given to illustrate this phenomenon. At present, very little is known about the regulatory processes that determine the level of expression of a particular pathway. It is clear, however, that for many products of secondary metabolism, an organized structure is required, indicating a regulatory hierarchy in which the morphology is a dominant factor. This is particularly apparent in the formation of, for example, secreted terpenes. Usually such hydrophobic compounds are accumulated in plant tissue in specialized glands, either within or on the surface of the plant organ. These glands also appear to be the site of biosynthesis. Cultures lacking such structures are not competent in the accumulation of these products. With other groups of compounds, an identifiable plant organ is required for accumulation to occur, such as the need for a root to obtain tropane alkaloids. In these examples, however, no clearly defined specialized structure is apparent within that organ, although there is some evidence that the biosynthetic capacity may be localized in specific cell types within the organ.

7.1 Production stability in cultures—effect of inducing morphology

Inducing morphological structure in culture can have two effects:

- to enable the accumulation of a group of compounds
- to alter the qualitative composition of the products accumulated

7.1.1 Mint oil production by *Mentha* cultures

Callus cultures of *Mentha* species lack any measurable amounts of mint oil terpenes, whether derived by phytohormone-induction or transformation. In contrast, when transformed galls of either *M. citrata* or *M. piperita* are induced to form shoots, the terpenes characteristic of the species appear in the culture. The formation of these compounds is directly correlated with, first, the appearance of leaf-bearing shoots and, secondly, with the presence on the leaves of oil-containing glands. When leaflets from these cultures are examined microscopically, the oil glands are seen to be identical to those on tissue from the parent plants.

7.1.2 Product profiles in cultures of *Ruta graveolens*

Aggregated suspension cultures of *R. graveolens* and shooty teratomas of this species, derived by transformation with *A. tumefaciens* CT58, all accumulate acridone and furoquinoline alkaloids and furanocoumarins (see Section 2.2). However, the two culture types show marked differences in both the constitutive levels of the different groups of compounds and their responses to elicitation. For example, the furanocoumarin, psoralen, is present at much higher levels in the shooty culture but is induced much more strongly in suspension cultures, reaching similar levels to those in the unelicited shoots. Also, the different compounds are elicited to various levels depending on the concentration of elicitor applied. Thus, different culture types can be used to study the responses of different chemical components to exogenous factors. All three chemical groups are derived from a common precursor, chorismic acid. Hence, the system is valuable for studying the way in which the secondary product groups interact with each other in relation to the supply of this common primary constituent.

7.2 Production stability in cultures—effect of disrupting morphology

If transformed root cultures are treated with exogenous hormones, the integrity of the organ is disrupted. With *D. stramonium* or *N. rustica* roots, treating with combinations of NAA (1–2 mg/litre) and kinetin (0.1–0.5 mg/litre) leads to the loss of the rooty phenotype. After about two to four days from application, elongation ceases and roots start to swell and thicken. Side branches start to fragment and an increasing number of small disorganized particles appear in the flask. Between about 6 and 14 days, these come to dominate the culture. When these small aggregates are passaged into fresh medium still containing the phytohormones, the culture continues to grow as a dispersed suspension. The total fresh mass accumulation is about comparable to that of the organized culture. Such cultures can be maintained for at least two years. On passage into phytohormone-free medium, some root primordia re-form. These can be picked out from the culture and grown on in phytohormone-free medium to yield regenerated root cultures.

If the alkaloid content of the culture is followed during this process, it is found that biosynthesis ceases rapidly in response to passage into phytohormone-containing medium. In *N. rustica*, the alkaloids appear simply to be diluted out by growth: in *D. stramonium*, in contrast, they appear to be actively degraded and largely lost from the system within 14 days.

Within 24 hours, alterations in the enzyme profile of the pathway can be seen, in particular in the activity of PMT. Up to 90% of the activity of this enzyme is lost by 12 hours and, although some recovery occurs, the activity is subsequently lost from the system and remains undetectable while cells are passaged in phytohormone-containing medium. Other enzymes are maintained at a low level. When the culture is returned to phytohormone-free medium, this activity reappears and, in parallel, the cultures return to the full level of competence in alkaloid formation. This phenomenon has been observed in a number of transformed root cultures. The extent to which reversion occurs is species dependent and probably also depends on the period of time spent in dispersed culture. This is discussed further in Section 7.3.

7.3 Chromosomal stability in cultures

Transformed roots usually show the normal level of ploidy for the species from which they are derived and retain this chromosomal complement over numerous years of culture. In contrast, dispersed or callus cultures show extensive aneuploidy and polyploidy, and the degree of genetic heterogeneity varies with time. When transformed tissue is grown as a dispersed culture, chromosomal aberration is introduced, leading to extensive aneuploidy in the population. *N. rustica* roots have been examined before, during, and after the establishment of a dispersed culture by treatment with phytohormone-containing medium. All integral 17-month-old roots show a chromosome number (n) of 48, as found *in planta*. In contrast, the range of n in a normal suspension culture of the same age is $16 \lesssim n \lesssim 82$. Eight weeks after passage into various media containing combinations of NAA, kinetin, and 2,4-D, only about 55% of cells examined show n = 48, the rest of the values ranging from $30 \lesssim n \lesssim 95$. After one year, less than 25% of cells had n = 48. When this one-year-old culture was passaged into phytohormone-free medium and the chromosome count of 85 ensuing root cultures examined, 70% showed n = 48, the majority of the others showing $40 \lesssim n \lesssim 50$. The growth rate and nicotine content of ten of these lines were examined and no clear correlation was found with the ploidy number of the roots. Interestingly, however, the aneuploid lines grew satisfactorily for at least one year and showed nicotine levels within the range of that found for roots of normal chromosome number.

8. Genetic manipulation of secondary product formation

For the genetic manipulation of secondary product pathways it is essential to have:

- the gene or genes to be manipulated
- a delivery system that will introduce the DNA into the genome of the desired plant species
- an expression cassette tailored to give expression of the gene in the right tissue and compartment
- a tissue culture and/or regeneration method in order that cultures and plants of the engineered strain may be obtained

It is beyond the scope of this chapter to give details of these aspects of genetic manipulation. Information on the approaches to identifying the important genes in the pathway is given in Section 6. Once the key genes have been identified, the proteins can be purified and used to clone the genes from cDNA libraries. Handbooks for these procedures are available, such as Deutscher (4) for protein methods, and Sambrook *et al.* (5) for DNA methods.

DNA can now be delivered to plants in a variety of ways:

- *Agrobacterium*-mediated plasmid-based integration (6)
- biolistic bombardment (7)
- uptake of naked DNA by protoplasts (see Chapter 3)
- ultrasonication (8)

The method of choice is probably still *Agrobacterium*-mediated plasmid-based integration but this cannot be used in many commercially important systems. A large range of expression vectors is now available (see ref. 6). These consist of a manipulated plasmid containing the T-DNA border regions surrounding a tailor-made expression cassette (*Figure 2*). The cassette will consist of a promoter region, a multiple cloning site, and a terminator sequence. The promoter may have upstream domains conferring tissue specificity or enhancing the strength of the promoter. The cloning site contains a number of unique restriction nuclease cleavage sequences, allowing the desired DNA to be inserted. Under the influence of the *vir* region of the *Agrobacterium* plasmid, the T-DNA, carrying the inserted gene, is transferred into the plant genome. Increasingly, a binary system is used, in which the foreign DNA is inserted into a cassette on a mini-plasmid which is placed into a host bacterium carrying a wild-type or disarmed plasmid. Excision and transfer of the foreign DNA takes place in *trans*. In either type of delivery system, a selectable marker gene is included in the T-DNA. Usually, this confers on the transgenic tissue resistance to an antibiotic (e.g. kanamycin, hygromycin) or a herbicide (e.g. bialophos). By growing transformants in the presence of the selecting agent, positive clones are identified.

Crops such as wheat, maize, and rice are monocotyledons and not susceptible to infection by *Agrobacterium*. Biolistics (7) have largely been developed to overcome this problem. In this method, DNA is coated on to the surface of minute particles of platinum or gold and fired into the tissue using a shotgun. A number of particles pass into or through cells, depositing DNA which, somehow, gets integrated into the nuclear genome. This method is also useful for non-integrative transient expression, where a quick analysis of constructs is required.

Often, especially in the secondary product field, the objective of genetic engineering is to increase flux into a particular product or group of products. In this case, it is desirable to introduce the target gene under a strong promoter, aiming to enhance the level of mRNA and of enzyme activity present in the synthetically active tissue. In other cases, however, it may be desirable to decrease the level of activity of an enzyme, thus down-regulating its activity. For example, it could be useful to produce strains capable of making only one of a range of products. This might be achieved by removing other enzyme activities. One approach to achieving this is to use antisense mRNA expression. The coding DNA of the target gene is inserted under a strong promoter in the reverse direction, forming an anti-mRNA. This appears to hybridize to the sense mRNA, preventing its translation to protein. An alternative technique, still being developed for use in plants, is to delete the target gene using site-directed mutation (9).

For manipulating a pathway in a plant, it is not always necessary to obtain the gene from the species of interest. Rather, a gene from another plant, an animal, or a micro-organism can be used, provided it is engineered correctly for expression in the plant system. Indeed, due to apparent negative inter-actions between the endogenous and inserted copies of homologous genes (10), it may in many cases be desirable to use a system of heterologous expression.

In the examples that follow, the production of a secondary product in culture has been influenced by the insertion of a foreign gene. Other ex-amples exist in which effects have been seen at the whole plant level.

8.1 Over-expression of a yeast ornithine decarboxylase gene in *N. rustica* root cultures

The biosynthesis of nicotine in tobacco uses putrescine, a diamine derived from either ornithine or arginine (*Figure 1*). Biochemical evidence has indicated that the supply of putrescine in root cultures might be limiting the ability of the culture to accumulate nicotine. The complete coding sequence of the gene for ODC from the yeast *Saccharomyces cerevisiae* is known, and the gene was therefore integrated into the binary vector pFIH10 (*Figure 2*) (11). This vector is designed to give high levels of constitutive expression, under the control of the enhanced CaMV 35S promoter. Similar constructs containing the chlor-amphenicol *O*-acetyl transferase (CAT) gene were used as control.

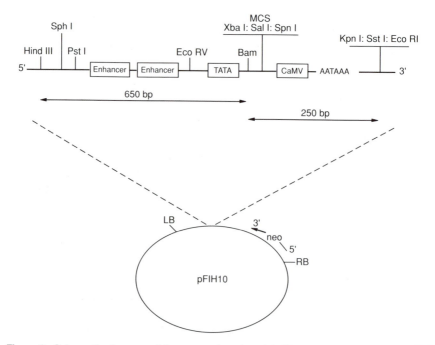

Figure 2. Schematic diagram of the expression plasmid pFIH10. MCS = molecular cloning site; RB = right border of T-DNA; LB = left border of T-DNA; neo = kanamycin resistance gene.

A number of cloned lines of transformed roots were examined for the levels of mRNA from the inserted ODC and for the level of expression of the enzyme. It was found that the mRNA was constitutively present throughout the growth cycle of the roots and that, in some lines, elevated amounts of enzyme activity also occurred. When the intermediates of the pathway and the end product, nicotine, were examined, some elevation was seen, particularly in the amount of nicotine accumulating by midgrowth phase. Thus, control lines transformed with pFIH10–CAT showed nicotine levels of 2.28 ± 0.22 μmol/g FM (N = 14), whereas lines transformed with pFIH10–ODC showed nicotine levels of 4.04 ± 0.48 μmol/g FM (N = 6).

Protocol 13. The insertion of the ODC gene into *N. rustica* roots and the selection of positive transformants

1. Excise the ODC structural gene from the plasmid pM13MP10 (11) as a *Hind*III : *Bsm*I fragment.

2. Clone into the *Hind*III : *Sph*I site of pMTL24 (11).

3. Isolate the coding sequence of the ODC gene as a 1.65 kb *Xba*I fragment.

4. Introduce this *Xba*I fragment into the expression vector pFIH10.

5. Transfer to *A. rhizogenes* LBA9402 by triparental mating, using *E. coli* containing the plasmid pRK2013 as the helper strain.

6. Select recombinants on YMB agar medium containing 100 μg/ml rifampicin and 50 μg/ml kanamycin sulfate.

7. Confirm presence of both plasmids and orientation of the ODC by Southern blotting or PCR.

8. Grow up *A. rhizogenes* in liquid YMB medium containing 100 μg/ml rifampicin and 50 μg/ml kanamycin sulfate.

9. Infect seedlings of *N. rustica* as described in *Protocol 5*.

10. Select transformed root cultures containing the ODC insert by culturing single root-tips on medium supplemented with 25–100 μg/ml kanamycin sulfate.

11. Establish a number of cloned lines.

12. Confirm insertion of the ODC by Southern blotting or PCR of the plant DNA.

13. Examine lines for levels of ODC mRNA and enzyme activity (see ref. 11).

8.2 Over-expression of a tryptophan decarboxylase gene in *Peganum harmala*

With the work described in Section 8.1, the effect of inserting the ODC gene on the amount of nicotine accumulated was small, only about twofold. This was presumably due to other enzymes becoming limiting once the block to putrescine formation was removed.

That this occurs has recently been demonstrated by inserting the gene for tryptophan decarboxylase (TDC) into *P. harmala* suspension or transformed root cultures (12). This enzyme catalyses the formation of tryptamine from tryptophan, the first step in the formation of serotonin, which then accumulates in these cultures. The pathway is very short, serotonin formation involving simply the 5-hydroxylation of tryptamine. It had previously been found that non-producing suspension cultures lacked the decarboxylase but not the hydroxylase activity. Root cultures produced levels of serotonin of about 0.5% DM.

When the gene coding for TDC, isolated from *Catharanthus roseus*, was inserted into these cultures, it was found to be expressed effectively, giving a three- to fivefold increase in extractable TDC activity. The engineered cultures contained, on average, about tenfold higher levels of serotonin, showing that the limitation in serotonin accumulation in the control tissue probably lay in the deficiency in TDC activity. Moreover, the authors were able to show that, in the engineered strains, serotonin accumulation was now limited by the availability of endogenous tryptophan. Exogenous tryptophan was rapidly metabolized to serotonin, about 30% being used in five days. In contrast, control lines used only about 2% in this time.

This work demonstrates very effectively the power of the approach of genetic manipulation. It also highlights an important issue—that the endogenous supply of the primary metabolite (in this case tryptophan) can become limiting when the engineering puts a high demand on the productive capacity of the plant. It confirms that, in some cases, it may be necessary to alter the regulation or the flux capacity of the primary metabolic pathways associated with the formation of the secondary product in order to realize the maximal potential of the latter. It should also be noted that, in the engineered *P. harmala* cultures, there was no significant increase in another group of tryptamine-derived secondary metabolites, the β-carbolines.

8.3 Over-expression of a lysine decarboxylase gene in *N. tabacum*

An added complication to genetic manipulation of secondary product formation in plants or cultures in illustrated in the work of Herminghaus *et al.* (13). These authors had, as their objective, the engineering of quinolizidine alkaloids, made in many members of the Fabaceae. The first step in this pathway involves the decarboxylation of lysine to cadaverine by lysine decarboxylase (LDC). This enzyme is present at rather low levels and, as with the other decarboxylases already discussed, might limit flux capacity into these products. A LDC gene (*ldc*) was isolated from the bacterium *Hafnia alvei* and introduced into *N. tabacum* cultures using *A. tumefaciens*. Cultures and plants were regenerated and the LDC protein and activity levels examined.

Initially, *ldc* was expressed under the control of the Tr-promoter and integrated, with the *neoII* gene for kanamycin resistance, into *A. tumefaciens* strain 50.154. Co-cultivation of tobacco protoplasts with bacteria yielded 15 kanamycin-resistant transformed calli. However, although in these calli *ldc* was clearly integrated and appropriate mRNA could be detected, LDC protein and activity could not be found in any of the calli tested.

It has been suggested that quinolizidine alkaloids are synthesized in the chloroplast. Therefore, a new construct was made in which *ldc* was linked to the promoter of the small subunit of Rubisco from potato. This construct, plasmid p22.154, was introduced into *A. tumefaciens* LBA4404 via triparental mating. Transformation of leaf discs of tobacco with this construct yielded 27 kanamycin-resistant plants, 15 of which showed mRNA for LDC in the leaves. LDC activity ranged from background up to 144 pkat/mg protein in a few examples. It appeared that LDC activity was located in the chloroplasts and it could not be detected in calli or roots. The over-expression of this enzyme had a dramatic effect on the cadaverine content of the leaves of transformed tobacco plants, increasing the level from barely detectable to as high as 1% dry mass. As this cadaverine was confined to the leaves, however, no influence on the alkaloid spectrum of the plants was seen.

Acknowledgements

I am most grateful to colleagues who have made available more details of their protocols than are available in the literature. Dr M. J. C. Rhodes kindly read the manuscript and made many valuable comments.

References

1. Kurz, W. G. W. (ed.) (1989). *Primary and secondary metabolism of plant cell cultures II*. Springer-Verlag, Berlin.
2. Rhodes, M. J. C., Spencer, A., Hamill, J. D., and Robins, R. J. (1992). In *Bioformation of flavours* (ed. R. L. S. Patterson, B. V. Charlwood, G. MacLeod, and A. A. Williams), pp. 42–64. Royal Society of Chemistry, London.
3. Zambryski, P. C. (1992). *Annu. Rev. Plant Physiol. Plant Mol. Biol.*, **43**, 465.
4. Deutscher, M. (ed.) (1990). *Guide to protein purification. Methods Enzymol.*, **182**. Academic Press, Orlando.
5. Sambrook, J., Fritsch, E. F., and Maniatis, T. (ed.) (1989). *Molecular cloning: a laboratory manual*. Cold Spring Harbor Laboratory Press, New York.
6. Jones, J. D. G., Shlumukov, L., Carland, F., English, J., Scofield, S. R., Bishop, G. J., and Harrison, K. (1992). *Transgenic Res.*, **1**, 285.
7. Batty, N. P. and Evans, J. M. (1992). *Transgenic Res.*, **1**, 107.
8. Joersbo, M. and Brunstedt, J. (1992). *Physiol. Plant.*, **85**, 230.

9. Offringa, R., van den Elzen, P. J. M., and Hookyaas, P. J. J. (1992). *Transgenic Res.*, **1**, 114.
10. Jorgensen, R. (1991). *TIBTECH*, **9**, 266.
11. Hamill, J. D., Robins, R. J., Parr, A. J., Evans, D. M., Furze, J. M., and Rhodes, M. J. C. (1990). *Plant Mol. Biol.*, **15**, 27.
12. Berlin, J., Rügenhagen, C., Dietze, P., Fecker, L., Goddijn, O. J. M., and Hoge, H. C. (1993). *Transgenic Res.*, **2**, 336.
13. Herminghaus, S., Schreier, P. H., McCarthy, J. E. G., Landsmann, J., Bottermann, J., and Berlin, J. (1991). *Plant Mol. Biol.*, **17**, 475.

9

Secondary products from cultured cells and organs: II. Large scale culture

A. H. SCRAGG

1. Introduction

Higher plants are still a major source of pharmaceuticals, colours, dyes, and flavours. Most of these plants are grown on large scale plantations, but it would be of value, particularly to the pharmaceutical industry, to be able to produce these compounds in a factory setting (1). Cultured plant cells were developed as a tool for studying plant metabolism, physiology, and development. However, plant cell cultures have also been seen as a possible source of natural products, and a large number of reviews have been written discussing the biotechnological potential of plant cell cultures (2–4). Most of these reviews have concentrated on pharmaceuticals with a high value such as those shown in *Table 1*, to which can be added a few more recent targets (5, 6). Many natural products have been detected in plant cell cultures, either callus or suspension, but only a few have been at levels which would make an industrial process possible (*Table 2*).

Inspection of *Tables 1* and *2* reveals that only a limited number of compounds are produced in useful amounts. This probably reflects our lack of knowledge of the pathways and controls of the synthesis of many natural products. One notable success was announced in 1983, the commercial production of shikonin from a culture of *Lithospermum erythrohizon* by Mitsui Petrochemical Company (7).

The production of natural products is not the only biotechnological use to which plant cells can be put, particularly in view of recent advances. Overall their potential industrial uses can be summarized as follows:

- production of natural or secondary products, pharmaceuticals, colours, dyes, or flavours
- biotransformations
- enzyme production

Table 1. Potential commercial products from plant cell cultures[a]

Product	Application	Plant source	Price ($/kg)
Ajmalicine (serpentine)	Anti-hypertensive	*Catharanthus roseus*	1500
Artemisinin	Anti-malarial	*Artemisia annua*	N/A[b]
Castanospermine	Glycosidase inhibitor, anti-HIV activity	*Castanospermum australe*	N/A
Codeine	Sedative	*Papaver somniferum*	1250
Digoxin	Heart stimulant	*Digitalis lanata*	3000
Diosgenin	Steroid	*Dioscorea deltoidea*	674
Guigkolides	Anti-inflammatory, PAF[c] antagonists	*Ginko biloba*	N/A
Jasmine	Fragrance	*Jasminum*	5000
Quinine	Anti-malarial	*Chinchona ledgeriana*	100
Shikonin	Anti-bacterial, dye	*Lithospermum crythrorhizon*	4500
Taxol	Anti-cancer	*Taxus brevifolia*	600 000
Vanillin	Flavour	*Vanilla planifolia*	1000
Vincristine Vinblastine	Anti-leukaemic	*Catharanthus roseus*	20 000 000
Yeuhchukene	Contraceptive	*Murraya paniculata*	N/A

[a] Data taken from refs 1, 5, 13, and 14.
[b] Not available.
[c] PAF, platelet aggregation factor.

Table 2. Levels of secondary products found in high producing plant cell cultures

Species	Product	Yield (% dry weight)
Coleus blumei	Rosmarinic acid	21.4
Morinda citrifolia	Anthraquinone	18.0
Coptis japonica	Benzlisoquinolines	15.0
Syringa vulgaris	Acetoside	15.0
Lithospermum erythrorhizon	Shikonin	12.4
Berberis wilsonae	Berberine alkaloids	10.0
Galium mollugo	Shikimic acid	10.0
Dioscorea deltoidea	Diosgenin	7.8
Nicotinana tabacum	Nicotine	5.0
Catharanthus roseus	Serpentine	2.2
	ajmalicine	1.8

- production of human protein via genetic engineering
- micropropagation

Whatever system is used the culture will need to be grown in volumes which will require the use of bioreactors. The type of culture grown will depend on the process, and this will in turn determine the type of bioreactor used. Before discussing bioreactor design and use, the various culture types are outlined.

1.1 Production of natural or secondary products

Systems developed for the production of secondary products have concentrated on suspension cultures, particularly in the early work, and these have proved to be easy to grow but difficult to manipulate for high yields (8). In addition to changes in cultural conditions, numerous methods have been used to stimulate product accumulation. Elicitation and precursor feeding have met with limited success (see *Tables 1* and *2*). Alternative approaches include:

- use of root, shoot, or embryo cultures
- use of 'hairy' root, transformed cultures (see Chapter 8)
- use of immobilized cultures

It has been observed that secondary product accumulation can often be initiated if the culture is allowed to differentiate into roots, shoots, or embryos (9). Immobilization of plant cells increases the production of secondary products in some cases (10) and has a number of other advantages over the use of suspension cultures:

- the biocatalyst is easily recovered and reused
- the product is easily separated from the catalyst
- increased stability of the biocatalyst
- continuous operation

1.2 Biotransformations

The nearly unlimited enzymatic potential of plant cell cultures can in principle be employed for bioconversions. The best known example is the conversion of β-methyldigitoxin to β-methyldigoxin by *Digitalis lanata* cells (11). Cell suspensions, immobilized cultures, and organized cultures can all be used for biotransformations.

1.3 Enzyme production

Plant cells contain a vast array of enzymes and both organized and suspension cultures could be a good source of these (see Chapter 8).

1.4 Production of human proteins

Agrobacterium tumefaciens has been used to transfer genes coding for antibodies to plants (12). In a similar way, human proteins have been produced by a cell culture of *Nicotiana tabacum* (13).

1.5 Micropropagation

Micropropagation via somatic embryogenesis, either directly or via callus, involves the production of a large number of adventitious embryos. This can occur at the callus stage or in liquid culture (14). This method of regeneration is attractive as liquid culture allows the use of a bioreactor; the production of many embryos in such a vessel results in a reduction in handling which is a major cost factor. This would allow the micropropagation of a number of seed crops for which the technique has been too expensive to date, as well as the production of artificial seeds (see Chapter 6).

2. Properties of plant cell suspensions and organized cultures

The type of bioreactor used and its design is dependent on the properties of the cultures used. Plant cell suspensions were first grown in bioreactors in the 1960s using various commercial or non-commercial designs adapted from animal cell culture. These designs are described in a review by Martin (15). Commercially produced stirred-tank bioreactors (STR) were initially used (*Figure 1*), in particular in Japan for the cultivation of tobacco cells (*N. tabacum*). Volumes up to 6500 litres were cultivated using low impeller speeds (below 100 r.p.m) as the cultures were considered to be sensitive to shear. The 1970s saw the introduction of the airlift bioreactor as an alternative to the stirred-tank because of its low shear characteristics (*Figure 1*).

Table 3 lists some of the characteristics of suspension and organized cultures. As the bulk of the research with bioreactors has used suspension cultures, this chapter will concentrate on these and discuss later the modifications used to cultivate 'hairy' roots and embryos.

Unlike microbial suspension cultures plant cell suspensions generally consist of groups or aggregates of many thousands of cells and up to 2 mm or more in diameter (*Figure 2*). A fine suspension would have aggregates in the range of 200–500 μm in diameter. The cells making up the aggregates are

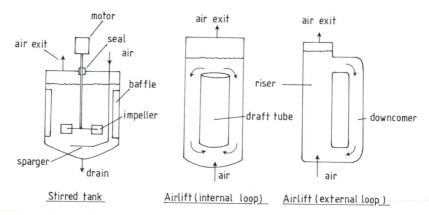

Stirred tank   Airlift (internal loop)   Airlift (external loop)

Figure 1. Two main designs of bioreactor; the stirred-tank (STR) and the airlift (internal draft tube or external loop).

Table 3. Characteristics of plant cell suspensions and organized cultures

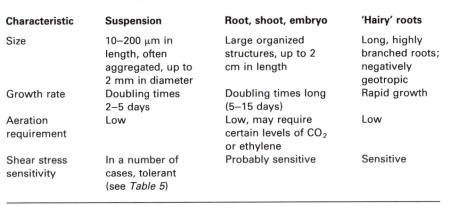

Characteristic	Suspension	Root, shoot, embryo	'Hairy' roots
Size	10–200 μm in length, often aggregated, up to 2 mm in diameter	Large organized structures, up to 2 cm in length	Long, highly branched roots; negatively geotropic
Growth rate	Doubling times 2–5 days	Doubling times long (5–15 days)	Rapid growth
Aeration requirement	Low	Low, may require certain levels of CO_2 or ethylene	Low
Shear stress sensitivity	In a number of cases, tolerant (see *Table 5*)	Probably sensitive	Sensitive

large, in the range of 40–200 μm in length. As a consequence of these properties the culture will settle out rapidly if mixing is stopped and sampling can therefore be difficult. The cultures grow very slowly, compared with microbial cultures, with doubling times of two to six days. This means that the bioreactor runs will be long, two to three weeks, but the aeration requirements are low at 1–10 mmol O_2 litre^{-1} h^{-1}, some tenfold less than for most micro-organisms. Thus the plant suspension culture will require good mixing without excessive aeration, and the long culture period will necessitate care in maintaining sterility.

Two other important characteristics are foam production and pH. Many plant cell suspensions in bioreactors produce foam which can develop into a matrix which can trap cells and block air exits. This 'meringue' can be reduced by the addition of antifoam (silicon-based or polypropylene glycol), but this may affect the culture. Microbial cultures in bioreactors often have their pH value controlled between two set values. In contrast, plant cell suspensions have the ability to control their own pH and pH control is therefore not normally used.

Other features of plant cell suspensions of particular relevance to their culture in bioreactors are their shear sensitivity and viscosity at high densities.

2.1 Culture rheology

In order to increase the productivity of slow growing cultures, a high cell density would be an advantage. The maximum has been proposed at 60 g/litre (dry weight) (16) and using perfusion culture a level of 75 g/litre (dry weight) has been achieved with a culture of *Coptis japonica* (17). At these cell densities it has been suggested that the culture would be so viscous as to cause problems of mixing (18). There have only been a few studies on the rheological

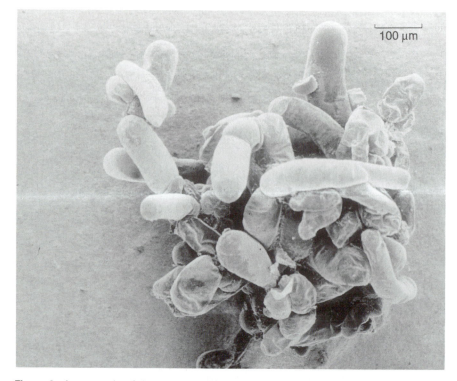

100 μm

Figure 2. An example of the aggregates in a tobacco cell suspension culture.

behaviour of plant cell cultures (see *Table 4*). The consensus is that plant cells are non-Newtonian and pseudoplastic but there is some discussion as to the presence of yield stress or shear thinning. This probably stems from the aggregated nature of the cultures, which causes rapid settling.

Viscosity can be measured using a number of techniques, but for cell suspensions the Brookfield-type rotational viscometer (cup and bob) has been widely used. In this system a cylinder is rotated at fixed speeds in a large vessel (infinite field) or in a vessel with a small and regular gap (*Figure 3*). While these types of viscometer are excellent for measuring fluids, giving reliable results, and rigorous measurements, they are not suitable for particulate suspensions. They suffer from phase separation due to settling out of particles, and those with a small gap can cause particle disruption. The cone and plate suffers from the same problems. These problems have been addressed by replacing the bob with a turbine impeller (*Figure 4*) in order to reduce settling and phase separation (24). A horseshoe-shaped impeller has been used to estimate the viscosity of a *Catharanthus roseus* suspension (23) (*Figure 3*). These alternative impellers do not, however, produce laminar

Table 4. Rheological properties of plant cell cultures

Culture	Measurement technique	Conclusions	Reference
Nicotiana tabacum	Brookfield rotational viscometer	High viscosity	19
Catharanthus roseus Morinda citrifolia	Brookfield rotational viscometer	Shear thinning thixotropic	20
Cudronia trianspidata Vinca rosea Agrostemma githagao	Brookfield rotational viscometer viscometer	Non-Newtonian pseudoplastic pseudoplastic	18
Catharanthus roseus	Horseshoe impeller	Yield stress Bingham and non-Newtonian	21
Catharanthus roseus	Contraves double-gap anchor impeller	Medium Newtonian, cells non-Newtonian; pseudoplastic 1–2 mPa/sec	22

flow which is required for absolute measurement of viscosity. They therefore cannot be used for the estimation of true viscosity.

2.2 Shear sensitivity of plant cell cultures

Plant cells in suspension have been regarded as sensitive to shear stress because of their rigid cellulose-based cell wall, relatively large size, and extensive vacuole, and this property has been used to explain the failure of plant cells to grow in some bioreactors. Reports concerning the shear sensitivity of plant cell suspensions are limited, but research has produced some very interesting results (25–28). The data indicate that although some cell lines are indeed sensitive to shear, many cultures are remarkably tolerant to shear levels (1000 r.p.m.; shear rate 167/sec) thought to be lethal (*Table 5*).

One of the major problems in the estimation of the effect of shear on viability is the measurement of viability itself. The viability of plant cells is often determined by membrane integrity using stain exclusion (Evans blue) or some metabolic activity rather than growth. The most commonly used metabolic activities used are the reduction of 2,3,5-triphenyltetrazolium chloride (TTC) (29) (see Chapter 7) and conversion of fluorescein diacetate (FDA) to fluorescein (30) (see Chapter 7). A method has been described whereby use of two radioactive compounds (glucose and mannitol), to one of which the cells are permeable and the other impermeable, allows the total intracellular volume of intact cells to be measured and hence an estimate of viable cells (31). The permittivity of cells at radio frequencies has also been used to determine the number of intact cells (28), and this has subsequently been used to follow the effect of shear on viability.

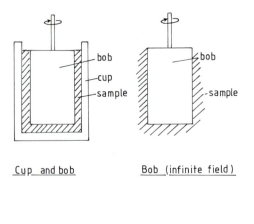

Cup and bob   Bob (infinite field)

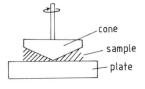

Cone and plate

Figure 3. Three forms of rotational viscometer; cup and bob, bob (infinite field), and cone and plate. In all three cases the viscosity of the sample is determined from the torque required to maintain a constant rotational speed of the bob or cone.

Table 5. The effect of shear on plant cell cultures

Cell line	Effect on viability
Short-term exposure	
(1000 r.p.m., 5 h)	
Catharanthus roseus (IDI)	Tolerant
Datura stramonium	Partial tolerance
Helianthus annuus	Tolerant
Vitis vinifera	Tolerant
Solanum tuberosum	Tolerant
Nicotiana tabacum	Sensitive
Picrasma quassioides	Tolerant
(5200 r.p.m. ultrathurax blender)	
Cinchona robusa	Tolerant
Nicotiana tabacum	Tolerant
Tabernaemontana divaricata	Sensitive
Long-term exposure	
(1000 r.p.m., 14 days)	
Catharanthus roseus	Tolerant
Nicotiana tabacum	Tolerant
Tabernaemontana divaricata	Sensitive
Cinchona robusta	Tolerant

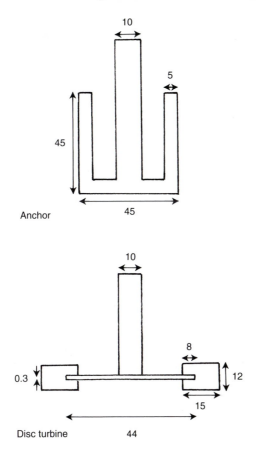

Figure 4. Alternative rotor designs for viscometers; anchor and disc turbine. All dimensions are in millimetres.

Although growth as a measure of viability can be misleading, because in some circumstances cells can remain viable but fail to grow, the ability of cells to grow can be used to determine the effect of shear (see *Protocol 1*). Other parameters such as enzyme release, oxygen uptake, and even FDA have proved difficult to use, probably due to the presence of cell debris in the suspensions. Also, the presence of an intact membrane is no guarantee of viability, as it is apparent that if a lethal stress is achieved, loss of viability is not instantaneous. Hooker *et al.* (26) have used TTC reduction, and permittivity (28) has also been used successfully to follow viability. Overall, the conclusions are that plant cell suspensions are more tolerant of shear than was previously expected, and this will perhaps lead to an increase in the use of stirred-tank bioreactors for cell suspensions.

Protocol 1. Determination of the effect of shear on plant cell culture

Equipment and reagents

- Sterile 3 litre stirred-tank bioreactor, diameter 15 cm, impeller diameter 7.3 cm
- Culture (2 litre) of *Catharanthus roseus* ILI in M3 medium (or your culture of choice)
- 24 sterile 250 ml flasks containing 100 ml of M3 medium (see *Protocol 2*)

Method

1. Set-up the 3 litre stirred-tank bioreactor for high r.p.m (1000 r.p.m) and with a suitable sampling device (see *Figure 5*).

2. Add 2 litres of 14-day-old *C. roseus* suspension under sterile conditions (if the bioreactor can be inoculated in a hood, do so).

3. Take a 50 ml sample, turn on the impeller motor, and adjust the speed to 1000 r.p.m.

4. Maintain the impeller speed at 1000 r.p.m and take 50 ml samples each hour for 5 h.

5. Take 2 × 20 ml aliquots from each sample and inoculate into 250 ml flasks containing M3 medium.[a] The remaining culture samples taken from the bioreactor (10 ml) should be used to determine wet and dry weights (using 3 × 3 ml) to determine any loss of weight during stress.

6. Incubate the 250 ml flasks under standard culture conditions (in this case, 25 °C, in diffuse light (1 μE·m^{-2}·s^{-1}), fluorescent light) at 150 r.p.m. on an orbital shaker).

7. Remove 3 × 3 ml samples from flask cultures at 0, 3, 6, 9, 12, and 15 days for wet and dry weight determination (see Chapter 1).

8. The growth data and loss of weight can be then used to estimate the effect of shear on the culture. Loss of wet and dry weight upon shearing will indicate cell breakage and reduced growth in terms of wet or dry weight will show any effects on viability.

[a] 20 ml is the standard inoculation volume for this *C. roseus* line which will grow well when diluted into 100 ml medium. If the inoculation density is a problem the samples can be filtered in sterile Hartley funnels and an equal wet weight (i.e. 3 g) used as an inoculum.

---

3. Bioreactor design

The two main bioreactor configurations are shown in *Figure 1*, the stirred-tank bioreactor design being the most common. Airlift bioreactors were introduced in the 1970s, particularly for biomass production, and were adopted for animal cell cultures because of their low shear characteristics.

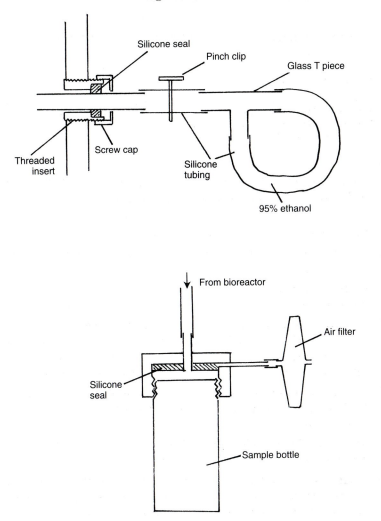

Figure 5. Designs for various sample ports for bioreactors cultivating plant cell suspensions.

3.1 Bioreactors for suspension cultures

Table 3 summarizes the main characteristics of plant cell suspensions which impact bioreactor design. Such cultures require good mixing, at moderate shear conditions, but have a low oxygen demand and thus require a lower K_La. This is in contrast to microbial cultures, where oxygen supply is often critical and mixing is needed for bubble break-up to increase oxygen supply.

With these constraints in mind, the airlift bioreactor has been used to grow a wide-range of plant cell suspensions. An example of the procedures involved is given in *Protocol 2*. This involves the checks which are required in order to ensure sterility, as plant cell cultures are particularly vulnerable due to their slow growth rate and thus need for extended bioreactor runs. These checks involve the testing of the reactor for leaks, the inoculum and medium for contamination, and a pre-inoculation incubation to determine whether the whole system is sterile.

Protocol 2. Culture of plant cells in an airlift bioreactor

The procedure used will greatly depend upon the bioreactor make and design, but the following outlines the general ideas which need to be borne in mind.

Equipment and reagents

- Bioreactor (30 litres airlift, L. H. Engineering)
- Inoculation culture, e.g. *C. roseus* IDI—the volume depends upon the bioreactor volume but 10% is usual, therefore 3 litres for a 30 litre vessel
- 30 litres M3 medium: MS medium (see Chapter 1) containing 1 mg/litre NAA, 0.1 mg/litre kinetin, and 2% (w/v) sucrose

Method

1. As 3 litres of inoculum will be required, start preparation two weeks before with the inoculation of five 1000 ml cultures in 2 litre flasks. Three flasks will be required to inoculate the vessel and the other two are spare in case of contamination or poor growth. If this is not done any contamination will set things back some four weeks.

2. If this is the first time the vessel has been used for plant cell suspensions give it a good clean and replace any seals or '0' rings which look worn. Remove the pH probe, acid and alkali addition lines, and seal the ports. Remove any other obstructions which could cause dead zones and trap cells. Replace the exit filter with a Buchner flask as a catch pot.

3. Fit the sample ports and enlarge if these are too small (see *Figure 5*).

4. A week before starting the culture, pressure test the bioreactor, especially if it is large (30 litres). Take it up to 1 bar empty and see whether this pressure can be maintained for 1 h. Any real loss of pressure will indicate a leak. It is not necessary to pressure test every time a run is to be made, but this check is advisable at regular intervals.

5. Three days before the start, fill the bioreactor with medium and sterilize *in situ* (for large bioreactors) according to the manufacturer's instructions. You can usually sterilize small (< 10 litre) bioreactors empty in an autoclave, although they can be sterilized containing medium. Remember it takes a considerable time to raise the centre of 10 litres of medium to 120 °C. A better alternative is to sterilize the medium in small volumes separately and to fill the bioreactor, preferably in a hood, after it has been sterilized. Often you may be able to fill a bioreactor with medium from a

---

large glass vessel which can be connected, using quick connectors, to the bioreactor.

6. Sterilize a sample of the medium separately, and retain for a sterility check.

7. Set the aeration level and temperature in the bioreactor to the values to be used during growth (normally 0.1–0.2 vvm, volume air/volume bioreactor/minute and 25 °C respectively) and leave the vessel for two or three days. The temperature controls will vary with the type of bioreactor as will the aeration system, but the latter should be controlled by a rotameter and pressure reduction valve, and the air passed through a 0.2 μm sterile filter. Sterilize these filters (of which there are numerous commercial forms) either with the bioreactor or separately and add later. Incubation under growth conditions will allow any contaminant present to grow and become visible. If the vessel has been fitted with a dissolved oxygen electrode (see manufacturer's instructions for use), any drop in dissolved oxygen will also indicate contamination.

8. Prior to inoculation, take a sample of the medium and inoculum for a sterility check. This consists of a microscopic observation, often a viability check (e.g. using FDA, see Chapters 1 and 7) and plating 0.1 ml on to YEPG plates where any contaminants will grow.
   These contain:
   • yeast extract (1% w/v)
   • peptone (1% w/v)
   • glucose (1% w/v)
   • agar (0.8% w/v)

9. If all the checks prove negative, inoculate the bioreactor with 3 litres of culture. In the case of the 30 litre airlift, inoculate using a large 5 litre flask connected to a large stainless steel needle (internal diameter 8 mm). Push the needle through a rubber washer in one of the top ports. Once the cells have been transferred, clip the tube off, and leave the flask in place.

10. If using a large bioreactor, remove culture samples each day for dry and wet weight determinations (see Chapter 1). Other parameters in the medium which can be followed include sugar, product accumulation, dissolved oxygen, and pH. Oxygen utilization and carbon dioxide evolution can be followed in the inlet and exit gas streams but it should be remembered that, as plant cells grow slowly, these change little.

11. Run the bioreactor until growth has ceased or the objective has been achieved. Reduce problems of foaming and meringue formation by the addition of antifoam or, for smaller reactors, by giving a daily shake!

The highly aggregated nature of the cultures and meringue formation can cause problems of sampling, non-homogeneous cultures, and loss of biomass. In

211

Figure 6. An example of meringue formation in a 7 litre airlift bioreactor containing a suspension culture of *Catharanthus roseus*.

such cases, modifications to the bioreactor should be made; for example, enlarging the sample ports and including some method for resuspending meringue.

Enlarge the sample ports to 8 mm (internal diameter) if possible. Two forms of sample port have been successfully used (see *Figure 5*). The first consists of a glass 'T' piece attached to the sample port and a silicon tubing loop between the other arms of the 'T' piece. The tubing loop is maintained sterile between samples by containing 95% ethanol. Upon sampling, run the ethanol to waste along with the culture medium in the tubing (i.e. not part of the culture itself); after this, run out the sample (20–50 ml) under gravity. The large volume of material taken with each sample means that this method should only be used with cultures of 5 litres or above. If the sample is required for determination of both wet and dry weights as well as product levels, volumes of 20–50 ml will be required. The sampling frequency should be one sample per day (doubling time of two to six days) unless the culture is very slow growing (doubling time longer than six days), in which case over 1 litre of culture will be removed over a period of two or three weeks. If the bioreactor is small (e.g. 2 litres) this will inevitably cause problems with airlifts, and will lead to changes in conditions in other bioreactors due to the lowering of the medium level. The second method of taking samples is that used with many microbial cultures, where blocking the air exit will force out culture under pressure. However, for plant cells the sample tube will require enlarging, otherwise the only sample obtained will be of culture medium.

Meringue formation occurs when the culture begins to foam due to accumulation of proteins in the medium (see *Figure 6*). Plant cell aggregates get trapped in this foam to produce a mixture of cells, polysaccharides, and proteins which can extend across the surface of the bioreactor, reach 10 cm or more in depth, and block exit and sample ports. This is a particular problem in airlift bioreactors with their high air input and high aspect ratio. Meringue formation can be inhibited by addition of antifoam agents, but these should be checked for inhibitory effects on the culture before use. Use of mechanical scrapers may reduce meringue, and for some small bioreactors a solid shake each day is often sufficient to dislodge meringue. For larger vessels (e.g. 20 litres) meringue does not appear to be a problem. In extreme cases of meringue formation the loss of biomass is very evident with a virtual clearing of the medium.

In spite of their advantages, airlifts have not been exclusively used, and plant cell suspensions continue to be grown in stirred-tank bioreactors; in fact, the two largest cultures to date (75 000 litres and 5000 litres) used this design. This probably reflects the general availability of the stirred-tank rather than a specific preference. Stirred-tank bioreactors can be modified to satisfy low oxygen and shear requirements (see *Protocol 3*).

Protocol 3. Culture of plant cells in a stirred-tank bioreactor

As the bioreactor operation will depend upon the manufacture and design of the bioreactor, only generalities can be given.

Equipment and reagents
- Stirred-tank bioreactor
- Inoculation culture

Method
1. As for the airlift bioreactor (*Protocol 2*), keep the system as simple as possible and free of obstructions. Therefore, if present, remove those connections and probe associated with pH control, but retain an anti-

---

Protocol 3. *Continued*

    foam addition port if this will be required during the bioreactor run. Often it is sufficient to add antifoam at the beginning along with the medium.

2. Remove the baffles and fit one of the alternative impellers (see *Figure 7*) if the culture is shear sensitive. Otherwise use the normal disc or flat bladed turbine at speeds between 100–300 r.p.m. depending on the culture.

3. If necessary, enlarge the sample ports or system as for the airlift bioreactor.

4. Remove the exit air filter and replace with a Buchner flask fitted with a cotton wool bung. This will collect any culture which foams over. If loss of water is a problem, fit a condenser on the exit air line. In many cases this is not required as the aeration rate is generally low for a stirred-tank bioreactor.

5. Set the aeration rate at 0.1–0.2 vvm (volume gas/volume bioreactor/ min). This is less than that used for the airlift bioreactor as the impeller is responsible for mixing in the stirred-tank bioreactor.

6. Prepare, inoculate, and run the bioreactor as described for the airlift bioreactor.

Most of the modifications have been associated with the impeller. Commercial bioreactors are generally fitted with a Rushton turbine impeller (see *Figure 7*), which gives good mixing and bubble break-up as it produces a turbulent flow with high shear. Early workers with plant cell cultures reduced the speed of rotation of these impellers from ∼ 1000 to ≤ 100 r.p.m. in order to reduce shear. Alternative impeller designs include anchor, inclined, helical, intermig, and large paddle (see *Figure 7*). All these provide mixing at lower shear rates than the turbine. All have been used with some success with a number of cultures, but to date no direct comparisons have been made and therefore no one design can be recommended. However, with the evidence that plant cell suspensions are less shear sensitive than was at first thought, the use of stirred-tanks fitted with a variety of impeller designs run at higher speeds (200–300 r.p.m.) may be of general use.

3.1.1 Alternative designs

Bioreactor designs other than the stirred-tank or airlift have also been proposed for the cultivation of plant cells. One approach has been to supply the oxygen for the culture indirectly rather than by sparging by using coils of silicone tubing; this material is permeable to oxygen and carbon dioxide (32). Two low shear mixing systems which have been used are the rotating drum bioreactor (33) and another with a rotating inner cylinder which mixes by Taylor vortises (34) (see *Figure 8*). Oxygen is supplied in the rotating drum by surface aeration and in the second bioreactor through a semi-permeable inner

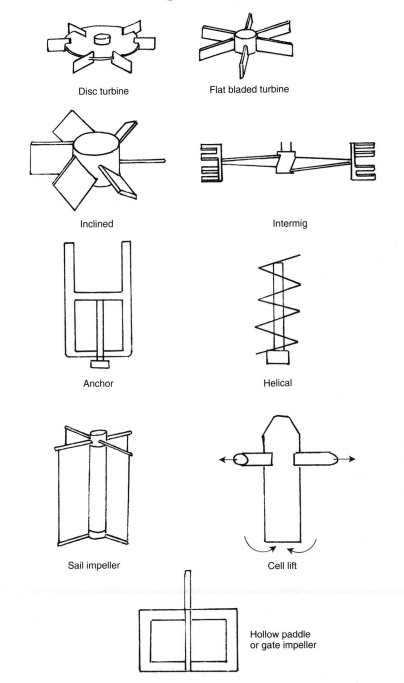

Disc turbine

Flat bladed turbine

Inclined

Intermig

Anchor

Helical

Sail impeller

Cell lift

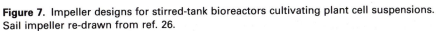

Hollow paddle
or gate impeller

Figure 7. Impeller designs for stirred-tank bioreactors cultivating plant cell suspensions. Sail impeller re-drawn from ref. 26.

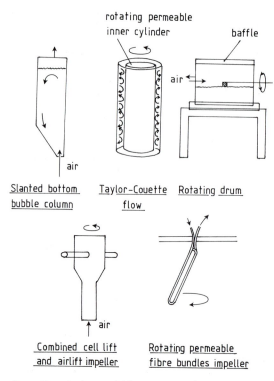

Figure 8. Three alternative designs of bioreactor and two impeller systems used to cultivate plant cell suspensions. The slanted bottom bubble column was re-drawn from ref. 60, the Taylor-Couette flow from ref. 34, the rotating drum from ref. 16 with the permission of Wiley, New York, the combined cell and airlift impeller from ref. 56, and the rotating permeable fibres from ref. 32.

drum. These new designs have had some success, indeed the rotating drum was shown to be effective (35), but all require specific construction as they are not manufactured.

3.2 Bioreactors for organized cultures

Organized cultures consist of roots, shoots, or embryos, which are large structures and are prone to damage due to their size. To date, no-one has determined their true shear sensitivity, but it is self-evident that they must be shear sensitive. Thus, if these cultures are to be grown in bioreactors, some form of low shear system will have to be used, and in the case of shoot cultures light may have to be provided. A number of bioreactors can be used for the growth of such cultures, and a list of these is given in *Table 6* and ref. 50. Most of these designs have been proposed for suspension cultures, except the spin filter and mist system (see *Figure 9*). In the spin filter system the spinning inner mesh cylinder allows removal of spent medium and its replace-

Table 6. Bioreactor designs used for organized cultures

Structure	Plant	Bioreactor design [a]	Reference
Embryo	*Daucus carota*	4 litre, STR	53
Embryo	*Medicago sativa*	1 litre spin filter	54
Roots	*Atropa belladonna*	10 litre modified STR [1]	55
Embryo	*Medicago sativa*	Airlift	54
Embryo	*Medicago sativa*	2 litre airlift [4]	56
		2 litre STR [5]	
		2 litre vibro-mixer [6]	
Plantlets	*Artemisia annua*	2 litre square airlift	51
Embryo	*Digitalis lanata*	5 litre airlift	52
Mini-corms	*Gladiolus*	Bubble column	50
Mini-tubers	*Solanum tuberosum*		
Plantlets	*Fragaris*		
Shoots	*Nephrolepis*		
	exaltata (fern)	1 litre bubble column	57
Embryo	*Daucus carota*	1 litre spin filter [2]	49
Embryo	*Panax ginseng*	Spin filter [2]	50
Embryo/shoots		Rotating drum	50
Embryo	*Euphorbia pulcherrima*		
	(poinsettia)	Silicone tubing	58
Shoots	*Musa* (banana),		
	Cordyline, Nephrolepis		
	(fern)	Mist bioreactor [3]	59

[a] Bioreactors [1], [2], and [3] are shown in *Figure 9*. Sources of the other bioreactors are: [4] L. H. Engineering Ltd., Stoke Poges, UK; [5] New Brunswick, Edison, USA; [6] Pegasus Industrial Specialities Ltd, Agincourt, Canada.

ment without removing the organized structure. Practically, it is difficult to take representative samples and maintain sterility once the culture has been initiated. Growth or development must be followed using indirect methods such as medium depletion (sugar use), conductivity (36), or permittivity (28).

3.3 Bioreactors for transformed cultures

'Hairy' root cultures grow as a tangled mass of branched roots which can be negatively geotropic, and as such behave somewhat like mycelial cultures. The long roots are sensitive and will be cut by an impeller. If this is too extensive the roots can revert to callus. It has been found that the cultures grow well if a support is provided in the bioreactor; this has the added advantage of separating the roots from the agitation system.

A number of support systems have been used, including stainless steel hooks (37), stainless steel mesh, polyurethane foam (38), and nylon mesh (39). The same problems of sampling exist as for organized cultures in that representative samples cannot easily be taken and therefore indirect methods are required. In one case the bioreactor had to be drained and the bioreactor plus biomass weighed at intervals (40).

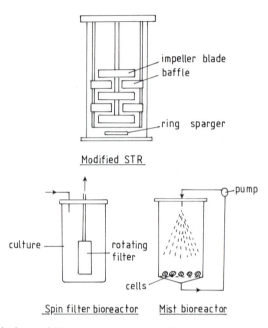

Figure 9. Three designs of bioreactors used to cultivate organized plant cultures. The modified STR was re-drawn from ref. 55 and the spin filter bioreactor re-drawn from ref. 49.

In the method described in *Protocol 4* the roots are immobilized in nylon mesh and growth followed by determination of sugar utilization.

Protocol 4. Growth of hairy roots in bioreactors

Hairy roots can be damaged in a stirred-tank bioreactor by the impeller, but by adding a mesh to separate the roots from the impeller a stirred-tank bioreactor can be used to grow these roots. Therefore, almost any design of stirred-tank bioreactor, airlift, or bubble column can be used.

Method

1. Ensure that the bioreactor is free of obstructions and as simple as possible. If an impeller is to be used install a mesh separating it from the culture. A number of meshes have been used; stainless steel, nylon, or polyurethane foam are all suitable. These can be attached to the baffles to replace the draft tube in internal loop bioreactors.

2. Do not enlarge the sample ports as it is not possible to take samples of the roots.

---

3. Use root segments as inoculum; these do not readily pass through tubing and hence inoculation may be difficult. If possible, use a large port to allow the root segments to be introduced into the vessel, and perform the inoculation in a hood.

3.4 Bioreactors for immobilized cultures

Immobilized plant cells have been shown to have a number of advantages over normal cultures. Immobilization allows continuous operation, separates biomass from medium, extends usable life, protects from shear, and appears in some cases to stimulate secondary product formation. The two main disadvantages are that zero growth or near zero growth conditions are needed, and that secondary products need to be exported to the medium. Given that these five criteria can be satisfied, there are a wide-range of bioreactors which can and have been used. The best design to use depends very much on how the cells have been immobilized. For plant cells the most popular methods of immobilization are:

- entrapment in polymers such as alginate, agar, carrageenan, polyurethane, or polyester
- entrapment behind a membrane
- attachment to a surface

Entrapment in polymers has mainly used alginate as this can be polymerized at room temperature using Ca^{2+} salts.

Protocol 5. Preparation of an immobilized plant cell bioreactor

Equipment and reagents

- Bioreactor: the design of bioreactor will depend on the method used to immobilize the cells and the process format used—these are outlined in *Figure 10* (this protocol uses a fluidized bed for alginate beads)
- 4% (w/v) sodium alginate in growth limiting medium, containing one tenth the normal phosphate concentration—autoclave this to form a solution

- 300 ml 0.2 M $CaCl_2$
- Lumbar puncture needle (diameter 2 mm)
- 25 ml sterile syringe
- Sterile beaker (one litre) and magnetic stirrer bar
- Growth limiting medium

Method

1. Add wet cells (filtered, usually about 16 g) to 100 ml of 4% sterile sodium alginate made up in medium, and mix. Carry out all manipulations under sterile conditions.

2. Fit a lumbar needle (2 mm) to the 25 ml syringe. (In some cases the needle and syringe can be replaced by a sterile burette.)

Protocol 5. *Continued*

3. Fill the syringe with the mixture.
4. Allow the mixture to drop into the CaCl$_2$ solution, which is being mixed using the magnetic stirrer. The whole system should be contained in a hood.
5. After the beads form, stir for 20 min in the CaCl$_2$ solution.
6. Decant the CaCl$_2$, wash once with medium, and suspend in growth limiting medium (200 ml).
7. Inoculate the bioreactor with the beads. In the case of a fluidized bed bioreactor, as shown in *Figure 10*, transfer a bead level of 2 g/10 ml to the bioreactor, i.e. 40 g wet beads.
8. Set the aeration rate at 75–100 ml/min using a rotameter and sterile air filter.

The polymer can be made into many shapes but beads are the most popular and easiest to make. Beads can be used in packed bed, fluidized bed, or airlift bioreactors (see *Figure 10*). Polyurethane generally comes in sheets (sound proofing), which can have a range of pore sizes, and can be easily cut into various shapes. It has been used in both packed and fluidized beds as cubes (0.8 cm) (41), shaped into a draft tube (42), or threaded as strips on stainless steel rods (43). Plant cells have been shown to attach spontaneously to glass, glass fibre mats (44), and non-woven short fibre polyester (45). As a high cell density is required to increase productivity, these materials have been wound on stainless steel supports within both sirred and airlift bioreactors (46).

Entrapment behind a membrane features in a number of bioreactor design options. Hollow fibre filtration units have been used (47), or simple flat plate systems (48). A more sophisticated bioreactor has been described where the membrane is a polypropylene sheet upon which cells are placed. Any growth of these cells is removed using a harvesting knife.

4. Bioreactor use

Once a culture is growing in a bioreactor, whatever the design, there are a number of formats in which the bioreactor can be used:

- batch
- continuous
- fed-batch
- draw-fill or semi-continuous
- perfusion

Normal batch growth is the most common format used whether accumulation of the product is growth linked or not. Continuous culture, which could

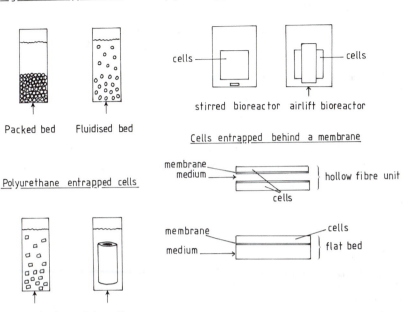

Figure 10. Some bioreactor configurations for use with immobilized plant cells: (A) alginate or polyurethane foam entrapped cells; (B) surface immobilized cells, re-drawn from ref. 45, and cells entrapped behind a semi-permeable membrane such as a hollow fibre unit or flat bed system.

theoretically increase productivity, has proved difficult with plant cell cultures. This has been due to the aggregated nature of the culture, meringue formation, and slow growth. The slow growth requires the supply of new medium at very low rates and a considerable period of time is therefore required to reach a steady state. This can be solved by the continuous supply of medium but only periodic removal of culture (27).

A fed-batch culture is one where a nutrient(s), usually sugar, is added in aliquots in order to avoid growth inhibition by using very high initial sugar concentrations. Draw-fill or semi-continuous cultures are those in which, at the end of growth, 80–90% of the culture is removed and replaced by fresh medium. In this way the lag phase is eliminated as such a high inoculum is used, and the bioreactors do not need sterilizing between runs. Perfusion culture is another method, like fed-batch, designed to increase the biomass levels and avoid inhibition. The spent medium is removed and new medium added behind a rotating mesh which excludes the cells. The mesh is rotated to avoid clogging. Another method has been to include a disengagement section which allows medium to be removed without cells.

5. Concluding remarks

From the preceding pages it should be clear that the bioreactor design and format used will depend upon the type of culture being grown. Therefore, no one design can be put forward as ideal; indeed very few comparisons have been made of one bioreactor with another using the same culture. However, the general principles behind the culture of plant cells in bioreactors are clear, and if the system is kept as simple as possible and great care taken with the maintenance of sterility, success should be achieved.

References

1. Berlin, J. (1984). *Endeavour*, **8** 5.
2. Barz, W. and Ellis, B. E. (1981). *Ber. Dtsch. Bot. Gaz.*, **94**, 1.
3. Berlin, J. (1988). In *Biotechnology in agriculture and forestry* (ed. Y. P. S. Bajaj), Vol. 4, pp. 37–51. Springer-Verlag, Berlin.
4. Fowler, M. W. and Stafford, A. M. (1992). In *Comprehensive biotechnology*, 2nd Supplement (ed. M. Moo-Young), pp. 79–85. Pergamon Press, Oxford.
5. Sahai, O. and Knuth, M. (1985). *Biotechnol. Prog.*, **1**, 1.
6. Walton, N. J. (1992). *Chemistry in Britain*, June, 525.
7. Curtin, M. E. (1983). *Bio/Technology*, **1**, 649.
8. Buitelaar, R. M. and Tramper, J. (1992). *J. Biotechnol.*, **23**, 111.
9. Charlwood, B. V., Charlwood, K. A., and Molina-Torres, J. (1990). In *Secondary products from plant tissue culture* (ed. B. V. Charlwood and M. J. C. Rhodes), pp. 167–200. Clarendon Press, Oxford.
10. Kurz, W. and Constabel, F. (1985). *CRC Crit. Rev. Biotechnol.*, **2**, 105.
11. Alfermann, A. W., Spieler, H., and Reinhard, E. (1985). In *Primary and secondary metabolism of plant cell cultures* (ed. B. Deus-Newmann, W. Barz, and E. Reinhard), pp. 316–22. Springer-Verlag, Berlin.
12. Hiatt, A., Cafferkey, R., and Bowdish, K. (1989). *Nature*, **342**, 76.
13. Hogue, R., Lee, J. M., and An, G. (1990). *Enzyme Microb. Technol.*, **12**, 53.
14. Sondahl, M. L. and Sharp, W. R. (1977). *Z. Pflanzenphysiol.*, **81**, 395.
15. Martin, S. M. (1980). In *Plant tissue culture as a source of biochemicals* (ed. E. J. Staba), pp. 149–66. CRC Press, Boca Raton, Florida.
16. Tanaka, H. (1987). *Process Biochem.*, Aug., 106.
17. Matsubara, K., Kitani, S., Yoshioka, T., Mommoto, T., and Fujita, Y. (1989). *J. Chem. Tech. Biotechnol.*, **46**, 61.
18. Tanaka, H. (1981). *Biotechnol. Bioeng.*, **23**, 1203.
19. Kato, K., Kuwazoe, S., and Soh, Y. (1978). *J. Ferment. Technol.*, **56**, 224.
20. Wagner, F. and Vogelmann, H. (1977). In *Plant tissue and its biotechnological applications* (ed. W. Barz, E. Reinhard, and M. H. Zenk), pp. 245–52. Springer-Verlag, Berlin.
21. Tanaka, H. (1982). *Biotechnol. Bioeng.*, **24**, 2591.
22. Scragg, A. H., Allan, E. J., Bond, P. A., and Smart, N. J. (1986). In *Secondary metabolism in plant cell cultures* (ed. P. Morris, A. H. Scragg, A. Stafford, and M. W. Fowler), pp. 178–94. Cambridge University Press, Cambridge.

23. Vogelmann, H. (1981). In *Advances in biotechnology* (ed. M. Moo-Young), pp. 117–21. Pergamon Press, Oxford.
24. Roels, J. A., van den Berg, J., and Voncken, R. M. (1974). *Biotechnol. Bioeng.*, **16**, 181.
25. Scragg, A. H., Allan, E. J., and Leckie, F. (1988). *Enzyme Microb. Technol.*, **12**, 362.
26. Hooker, B. S., Lee, J. M., and An, G. (1989). *Enzyme Microb. Technol.*, **11**, 484.
27. Meijer, J. J. (1989). PhD thesis, University of Delft, The Netherlands.
28. Markx, G. H., Ken Hooper, H. J. G., Meiger, J. J., and Vinke, K. L. (1991). *J. Biotechnol.*, **19**, 145.
29. Towill, L. E. and Mazur, P. (1975). *Can. J. Bot.*, **53**, 1097.
30. Widholm, J. M. (1972). *Stain Technol.*, **47**, 189.
31. Parr, A. J., Smith, J. I., Robbins, R. J., and Rhodes, M. J. C. (1984). *Plant Cell Rep.*, **3**, 161.
32. Piehl, G. W., Berlin, J., Mollenschott, G., and Lehmann, J. (1988). *Appl. Microb. Biotechnol.*, **19**, 456.
33. Tanaka, H., Nishijima, F., Suwa, M., and Iwanoto, T. (1983). *Biotechnol. Bioeng.*, **25**, 2359.
34. Janes, D. A., Thomas, N. H., and Callow, J. A. (1987). *Biotechnol. Tech.*, **1**, 257.
35. Shibasak, N., Hirose, K., Yonemoto, T., and Tadaki, T. (1992). *J. Chem. Tech. Biotechnol.*, **53**, 359.
36. Taya, M., Yoyama, A., Kondo, O., Kobayashi, T., and Matsui, C. (1989). *J. Chem. Eng.* Japan, **22**, 84.
37. Wilson, P. D. G., Hilton, M. G., Meehan, P. T. H., Waspe, C. R., and Rhodes, M. J. C. (1990). In *Progress in plant cellular and molecular biology* (ed. H. J. J. Nijkamp, L. H. W. van der Plas, and J. van Aartijk), pp. 700–5. Kluwer, Dordrecht.
38. Taya, M., Hegglin, M., Prenosil, J. E., and Bourne, J. R. (1989). *Enzyme Microb. Technol.*, **11**, 170.
39. Rodriguez-Mendiola, M. A., Stafford, A., Cresswell, R., and Arias-Castro, C. (1991). *Enzyme Microb. Technol.*, **13**, 697.
40. Wilson, P. D. G., Hilton, M. G., Robbins, R. J., and Rhodes, M. J. C. (1987). In *Bioreactors and biotransformations* (ed. G. W. Moody and P. B. Baker), pp. 38–51. Elsevier, London.
41. Mavituna, F. and Park, J. (1985). *Biotechnol. Lett.*, **7**, 637.
42. Ziyad-Mohamed, M. T. and Scragg, A. H. (1990). In *Methods in molecular biology* (ed. J. W. Pollard and J. M. Walker), Vol. 6, pp. 525–36. Humana Press, Clifton, NJ.
43. Robertson, G. H., Doyle, L. R., Sheng, P., Paulath, A. E., and Goodman A. (1989). *Biotechnol. Bioeng.*, **34**, 1114.
44. Facchini, P. J. and Di Cosmo, F. (1991). *Biotechnol. Bioeng.*, **37**, 397.
45. Archambault, J., Voleskey, B., and Kurz, W. G. W. (1990). *Biotechnol. Bioeng.*, **35**, 702.
46. Tom, R., Jardin, B., Chavarie, C., Rho, D., and Archambault, J. (1991). *J. Biotechnol.*, **21**, 21.
47. Prenosil, J. E. (1983). *Enzyme Microb. Technol.*, **5**, 323.
48. Shuler, M., Hallsby, G., Payne, G., and Cho, T. (1986). *Ann. N.Y. Acad. Sci.*, **469**, 270.

49. Styer, D. J. (1985). In *Tissue culture in forestry and agriculture* (ed. R. R. Henke and K. W. Hughes), pp. 117–30. Plenum Press, New York.
50. Takayama, S. (1992). In *Biotechnology in agriculture and forestry* (ed. Y. P. S. Bajaj), Vol. 17, pp. 17–34. Springer-Verlag, Berlin.
51. Park, J. M., Hu, W-S., and Staba, E. J. (1989). *Biotechnol. Bioeng.*, **34**, 1209.
52. Greidziak, N., Diettrich, B., and Luckner, M. (1990). *Planta Med.*, **56**, 175.
53. Kessell, R. H. J. and Carr, A. H. (1972). *J. Exp. Bot.*, **23**, 996.
54. Stuart, D. A., Strickland, S. G., and Walker, K. A. (1987). *Hortic. Sci.*, **22**, 800.
55. Akita, M. and Takayama, S. (1988). *Acta Hortic.*, **230**, 55.
56. Chen, T. H. H., Thompson, B. G., and Gerson, D. F. (1987). *J. Ferment. Technol.*, **65**, 353.
57. Ziu, M. and Hodar, A. (1991). *J. Bot.* (Israel), **40**, 7.
58. Preil, W., Florek, P., Wix, U., and Beck, A. (1988). *Acta Hortic.*, **226**, 99.
59. Weathers, P. J., Chedetham, R. D., and Giles, K. L. (1988). *Acta Hortic.*, **230**, 39.
60. Taticek, R. A., Moo-Young, M., and Legge, R. L. (1990). *Appl. Microbiol. Biotechnol.*, **33**, 280.

A1

Addresses of suppliers

Baekon Inc., 4220 Enterprise Street, Fremont, CA 94538, USA.

BBL, Becton Dickson and Co., PO Box 243, Cokeyville, MD 21030, USA.

BDH Chemicals, Broom Road, Poole, Dorset BH12 4NN, UK.

Bio-Rad Laboratories, 2000 Alfred Nobel Drive, Hercules, CA 94547, USA; Bio-Rad House, Maylands Ave., Hemel Hempstead, Hertfordshire HP2 7TD, UK.

Boehringer-Mannheim, 9115 Hague Road, Indianapolis, IN 46250–0414, USA; Bell Lane, Lewes, Sussex BN7 1LG, UK.

BTX, 11199 Sorrento Valley Road, San Diego, CA 92121–1334, USA.

Calbiochem, 10933 N. Torrey Pines Road, La Jolla, CA 92112–4180, USA; 3 Heathcoat Building, Highfields Science Park, University Boulevard, Nottingham NG7 2QJ, UK.

Carlo Erba/Fisons, 24911 Avenue, Stanford, Valencia, CA 91355, USA; via C Imbonati 24, I-20159 Milan, Italy.

Difco Laboratories, PO Box 331058, Detroit, MI 48232–7058, USA.

Falcon, (Becton Dickson Labware), Franklin Lakes, NJ 07417, USA.

Fisher Scientific, 711 Forbes Avenue, Pittsburgh, PA 15219–4785, USA; New Enterprise House, St. Helen's Street, Derby DE1 3GY, UK.

Fluka Chemicals, Ltd., The Old Brickyard, New Road, Gillingham, Dorset, UK.

FMC Bioproducts, 191 Thomaston Street, Rockland, ME, USA.

Gelman Sciences, see Fisher Scientific; 10 Harrowden Road, Brackmills, Northampton NN4 0EZ, UK.

Genecor International Inc., 1870–4 S. Winton Road, Rochester, NY 14618, USA.

GIBCO BRL, PO Box 68, Grand Island, NY 14072–0068, USA.

Hakko, Kyowa Kogya Co., Ltd., Ohtemachi Building, Ohtemachi, Ciyoda-Ku, Tokyo, Japan.

Hewlet Packard Co., HP Analytical Direct, 221 Gale Lane, Kennett Square, PA 19348, USA.

ICN Biomedicals Ltd., Eagle House, Peregrine Business Park, Gomm Road, High Wycombe, Buckinghamshire, UK.

L.H. Engineering, Stoke Poges, UK.

Magenta, 3800 N. Malwonkoe Ave., Chicago, IL 60341, USA.

Meiji Seika Kaisha Ltd., International Division, 4–16 Kyobashi 2-Chrome, Chuo-ku, Tokyo 104, Japan.

Merck, Reagents Division, Frankfurter St. 250, Postfach 4119, D-6100 Darmstadt 1, Germany.

Nalgene Co., 75 Panorama Creek Drive, Rochester, NY 14602–0365, USA.

New Brunswick Scientific Co., Inc., 44 Talmadge Road, Edison, NJ 08818–4005, USA.

Onozuka, (Yakult Honsha Ltd. and Serva Biochemicals).

Oxoid Ltd., Wade Road, Basingstoke, Hants RG24 0PW, UK.

Pharmacia LKB Biotechnology, 800 Centenial Avenue, Piscataway, NJ 08855–1327, USA; Davy Avenue, Knowlhill, Milton Keynes MK5 8PH, UK.

Prototype Design Services, 23 N. Pickley Street, Madison, WI, USA.

Serva Biochemicals, 200 Shames Drive, Westbury, New York 11590, USA; D-6900 Heidelberg, Germany.

Sigma Chemical Co., 3050 Spruce Street, St. Louis, MO 63103, USA; Fancy Road, Poole, Dorset BH17 7NH, UK.

Unwin, R.W. & Co., Ltd., Prospect Place, Welwyn, Hertfordshire, UK.

Waters Chromatography Division, Millipore Corp., 34 Maple Street, Milford, MA 01757, USA; The Boulevard, Blackmoor Lane, Watford, Hertfordshire WD1 8YW, UK.

Whatman Labsales, 5285 N.E. Elam Young Parkway, Suite A-400, Hillsboro, OR 97124–9981, USA; Unit 1, Coldred Road, Parkwood, Maidstone, Kent ME15 9XN, UK.

Yakult Honsha Co. Ltd, Medicine Department, Enzyme Division, 1-1-19, Higashi-Shinbashi, Minatokv, Tokyo 105, Japan.

Index